GM-Crop Cultivation – Ecological Effects on a Landscape Scale

Theorie in der Ökologie
Herausgegeben von Broder Breckling

Band 17

Broder Breckling
Richard Verhoeven
(eds.)

GM-Crop Cultivation – Ecological Effects on a Landscape Scale

Proceedings of the Third
GMLS Conference 2012 in Bremen

Bibliographic Information published by the Deutsche Nationalbibliothek
The Deutsche Nationalbibliothek lists this publication in the Deutsche
Nationalbibliografie; detailed bibliographic data is available in the internet at
http://dnb.d-nb.de.

Cover illustration:
© Broder Breckling

Typesetting, organisation and layout:
Richard Verhoeven

Library of Congress Cataloging-in-Publication Data

GMLS Conference (3rd : 2012 : Bremen, Germany)
  Gm-crop cultivation : ecological effects on a landscape scale : proceedings
of the third GMLS Conference 2012 in Bremen / Broder Breckling, Richard
Verhoeven (eds.). — 1st ed.
   p. cm. — (Theorie in der Ökologie, ISSN 1615-374X ; Bd. 17)
Includes index.
ISBN 978-3-631-62870-6
  1. Transgenic plants—Ecology—Congresses. 2. Transgenic plants—Risk
assessment—Congresses. I. Breckling, Broder. II. Verhoeven, Richard.
III. Title. IV. Series: Theorie in der Ökologie ; Bd. 17.
SB123.57.G633 2012
632'.8—dc23

2013006371

ISSN 1615-374X
ISBN 978-3-631-62870-6
© Peter Lang GmbH
Internationaler Verlag der Wissenschaften
Frankfurt am Main 2013
All rights reserved.
PL Academic Research is an Imprint of Peter Lang GmbH.

All parts of this publication are protected by copyright. Any
utilisation outside the strict limits of the copyright law, without
the permission of the publisher, is forbidden and liable to
prosecution. This applies in particular to reproductions,
translations, microfilming, and storage and processing in
electronic retrieval systems.

www.peterlang.de

Contents

Foreword	9
Organising committee	10
Welcome Address by Maike Schaefer – Vice-chairperson of the parliament-group of Bündnis90/Die Grünen Bremen and speaker of environmental policy	11

Chapter I: Dispersal and hybridisation of GMO

Subspontaneous glyphosate-tolerant genetically engineered *Brassica napus* L. along Swiss railways
    Nicola Schoenenberger & Luigi D'Andrea
   17

Variation of maize pollen shedding in North Germany and its relevance for GMO-monitoring
    Frieder Hofmann, Ulrich Schlechtriemen, Ulrike Kuhn, Klaus-Peter Wittich, Wolfgang Koch, Steffi Ober, Rudolf Vögel & Mathias Otto
   19

Wind-mediated pollen dispersal of oilseed rape – an estimation using pollen traps
    Wieslawa Poplawska, Alina Liersch, Malgorzata Jedryczka, Joanna Kaczmarek, Joanna Wolko, Maria Ogrodowczyk & Iwona Bartkowiak-Broda
   26

Investigation of oilseed rape gene flow using erucic acid as biochemical marker
    Iwona Bartkowiak-Broda, Wieslawa Poplawska, Alina Liersch, Tadeusz Walkowski & Maria Ogrodowczyk
   34

The influence of volunteers and soil seed bank on the quality of oilseed rape seeds
    Alina Liersch, Joanna Wolko, Wieslawa Poplawska, Krystyna Krotka & Iwona Bartkowiak-Broda
   38

Pollen flow evaluation to implement a Quick Monitoring Index (QMI)
    Elena Balducci, Donatella Paffetti, Davide Travaglini, Stefano Biricolti, Francesca Bottalico, Silvia Fiorentini, Anna Buonamici, Francesca Donnarumma, Alessandro Materassi, Gianni Fasano, Lorenzo Chelazzi, Filippo Cimò, Isabella Colombini, Laura Bartalucci, Antonio Perfetti, Olga Mastroianni, Valeria Tomaselli, Simone Gorelli, Francesco Tonazzini & Cristina Vettori
   44

Can dwarfed Oilseed Rape (*Brassica napus* L.) measure up to tall cultivars?   53
    Jana Seeger, Broder Breckling & Juliane Filser

Chapter II: Landscape effects and agro-ecological interferences

Domestication, feral species and the importance of industrial agriculture to the future of plant diversity   59
    Cynthia Sagers, Meredith Schafer, Brett Murdoch, Jason Londo, Steven Travers & Peter Van de Water

Large scale and small scale approaches for assessing potential exposure of habitats and species neighbouring GM plant cultivation   61
    Frieder Graef, Anne Heyer, Sigrid Ehlert, Ulrich Stachow, Claudia Bethwell, Sarah Effertz, Klaus Henle & Birgit Winkel

Coexistence in Maize: Efficacy of non-GM border rows in reducing pollen-mediated gene flow   67
    Maren Langhof & Gerhard Rühl

Maize gene flow simulation for intensively used agrarian areas in Lower Saxony (Germany)   71
    Markus Ernsing, Broder Breckling, Hauke Reuter & Gunther Schmidt

Modelling potential maize hybridisation in northern Germany and implementation of a WebGIS application for GMO monitoring issues: Two aspects of the GeneRisk project   78
    Gunther Schmidt, Broder Breckling & Winfried Schröder

Chapter III: Long-term experience and sociological consequences

Impacts of genetically engineered crops on pesticide use in the U.S. – The first sixteen years   89
    Charles Benbrook

Farmer's choice of seeds in five regions under different levels of seed market concentration and GM crop adoption   91
    Rosa Binimelis, Angelika Hilbeck, Tamara Lebrecht, Rapahela Vogel & Jack A. Heinemann

Implications of GM crops in subsistence-based agricultural systems in Africa Denis W. Aheto, Thomas Bøhn, Broder Breckling, Johnnie Van den Berg, Lim Li Ching & Odd-Gunnar Wikmark	93
Co-existence challenges in small-scale farming when farmers share and save seeds Thomas Bøhn, Denis W. Aheto, Felix S. Mwangala, Inger Louise Bones, Christopher Simoloka, Ireen Mbeule, Odd-Gunnar Wikmark, Gunther Schmidt & Ignacio Chapela	104

Chapter IV: Causes and effects – Research perspectives and requirements

Teratogenesis by glyphosate based herbicides and other pesticides. Relationship with the retinoic acid pathway Andrés Carrasco	113
Human cell toxicity of pesticides associated to wide scale agricultural GMOs Robin Mesnage, Steeve Gress, Nicolas Defarge & Gilles-Eric Séralini	118
Effect of extreme climatic conditions on Bt toxin concentration in transgenic maize Miluse Trtikova, Matthias S. Meier & Angelika Hilbeck	121
Cry1Ab toxin quantification in MON 810 maize András Székács	123
Are frogs and toads affected by complementary herbicides of GM crops? Norman Wagner, Wolfram Reichenbecher, Hanka Teichmann, Beatrix Tappeser & Stefan Lötters	125
Transgenic evolution and ecology are proceeding Broder Breckling	130

Chapter V: Transdisciplinary contributions:
Comments by stakeholders and administrators

Establishment of an European data centre for PMM – what will be the best option? Wiebke Züghart	139

Official seed monitoring as a potential data source for GMO monitoring 141
    Hans-Georg Starck

Impact of GMOs on the beekeeping sector: Still neglected in research, widely 144
ignored in regulation, untapped potential for monitoring
    Walter Haefeker

Deficits in research funding for analysis of health and environmental risks of 148
GM plants – the example of Germany
    Martha Mertens

Research Policy & Independent Risk Research – Draft demands by civil society 152
organisations for German parliamentary elections in 2013
    Christoph Then

A publication forum for GMO research, initiated by the GMLS conference: 156
Implications of Cultivation and Monitoring of Genetically Modified Organisms
    Gunther Schmidt & Broder Breckling

Foreword

Currently within the European Union, there are 47 genetically modified (GM) crops that have been approved. These consist of 27 maize varieties, 8 cotton, 7 soybean, 3 oilseed rape and 1 variety of each sugar beet and starch potato. The vast majority of these crops received approval only for import. Some of these are on the market as food and feed or as additives to them. Only two GM-plants, one maize variety and one potato, are authorised for cultivation in the EU so far.

Although genetically modified organisms (GMO) are widely rejected by consumers, producers and distributors in Europe, there are numerous GM events that have applied for EU market authorisation. There is an ongoing controversy in science and among regulators whether GM crops are safe and what the requirements are to demonstrate the absence of relevant risks in a trustworthy and reliable way. The undesirable and adverse effects on human health and the environment cannot be excluded. The established conventional agricultural economy as well as organic agriculture could be at risk.

On June 14-15, 2012, more than 60 experts from science and administration met in Bremen for the third conference on Implications of GM-Crop cultivation at Large Spatial Scales – GMLS III. They discussed a variety of topics on ecological questions and non-target effects, socio-economy and coexistence, methodological aspects and modelling, authorisation and regulation of GMO.

This volume compiles 28 contributions to the GMLS III conference from Argentina, Ghana, the United States and from eight European countries. Selected contributions are published in the series *Implications of Cultivation and Monitoring of Genetically Modified Organisms* of the SpringerOpen Journal *Environmental Science Europe* that refers to the GMLS conferences, but also has additional contributions. Both the conferences and the publication series attract a high degree of interest, evident from the high download numbers of the articles.

The editors gratefully acknowledge funding of the *Gekko Foundation* and the *Fondation Charles Léopold Mayer pour le Progrès de l'Homme*. Without their support, the conference would not have been possible.

Broder Breckling & Richard Verhoeven

Organising committee of the conference

Broder Breckling, University of Bremen, University of Vechta
Hartmut Meyer, ENSSER
Gunther Schmidt, University of Vechta
Winfried Schröder, University of Vechta
Christoph Then, Testbiotech
Richard Verhoeven, University of Bremen
Wiebke Züghart, Federal Agency for Nature Conservation

Participants of the GMLS III Conference in Bremen

Maike Schaefer

Vice-chairperson of the parliament-group of Bündnis90/Die Grünen Bremen and speaker of environmental policy

Dear Ladies and Gentlemen, dear guests,
welcome to Bremen and welcome to the GMLS Conference on Implications of Genetically Modified Crop Cultivation at Large Spatial Scales.

I am very proud that this international conference is taking place here in Bremen for the 3rd time.

The conference is held at the House of Science, which is located right in the historical city centre, near the Cathedral the Market Place with the House of Parliament, City Hall, and the Chamber of Trade and Commerce.

When I had a look at the conference program, I was honestly surprised not only by the many presenters from all over the world, but also by the wide range of the different and interesting aspects of GMO-assessment. The Conference attempts to bring together leading scientific expertise to assess impacts of genetically modified organisms in the context of agricultural applications. And I will comment the relevance of the conference topics from a scientific but also a political perspective.

Bremen's long experience in risk assessment

When we politicians talk these days about risk assessment in Bremen it is more about a financial risk assessment, being a poor federal country. So we can state, for the public institutions as well as the private sector: a good considered risk assessment is absolutely necessary.

And of course this applies also to genetically modified organisms, to chemicals and to other technical approaches. Having worked for many years as a scientist in general ecology and ecotoxicology, I am well familiar with the requirements of environmental risk assessment. Compared to chemicals, GMO require a more comprehensive analysis. This includes not only a compositional analysis but also assessment of physiological performance, aspects of cultivation and ecological effects as changes in biodiversity. Monitoring is an additional task.

To make reasonable decisions, policy depends on an information basis that is well balanced. It is highly important, that not only the view of the decision makers is available. For a reasonable risk assessment, critical and independent research is indispensable. The public funding of scientific expertise that is not involved in a specific interest is a MUST for regulators if they want to be efficient.

And please allow me this statement – it would be even more helpful, if more scientists would spent at least some time of their carrier as politicians- because my experience over the last five years- since I am into politics- is, that we have a great lack of knowledge and environmental understanding within the group of decision makers. We have a great deal of relevant and helpful information published by scientists and researchers ... but when it comes to practical decisions in politics, we have a lack of experts there and a high number of non-experts, often more driven by the interests of their electors or of their party platform or lobbies, than by objective scientifically facts.

But I don't want to wallow in self-pity and pessimism. I will rather show you the optimistic sites:

GMO policy in Bremen

Let me inform you about some political decisions the Free Hanseatic City of Bremen has made. Bremen has decided that agricultural contractual partners of the city working on municipal areas are obliged to cultivate conventional varieties. Bremen also officially declared itself as a GMO-free district, which means that all farmers here committed themselves only to grow conventional crops, no GMO. In the general public, this policy has a considerable support.

As a politician, I have to emphasise, that in decision making on GMO, scientific information is highly important. Value-based consumer preferences and the protection of GMO free production are at least equally relevant. The value preference for food being as natural as possible has a very high priority in consumer decisions here in Germany. More and more people go for organic food.

I was actually contacted by the local bee-masters some months ago, who were eager to learn something about bees being at risk by GMO. I was really surprised and happy to see, that these men were seeking contact with politicians and researchers. And by chance only some weeks later, we had the judgement by the European Court that even the smallest traces of unapproved GMO found in honey, means that this honey may not go into the shops. And that means all the concerned bee-masters have the right to bring their case to trial. In my eyes, this was a great success.

I expect that the conference results and its documentation will be appropriately considered on the scientific, on the public and on the administrative level. For this conference, the organising committee has brought together contributions from Europe and overseas which provide important new insight and experiences. So I expect a relevant impact in the discussion on GMO and the regulation of the involved risks.

I hope, that your stay in Bremen offers useful scientific information, that you make useful contacts, and it will bring further inspirations for your personal work.

Enjoy your stay in Bremen. Take some time to enjoy the highlights in the city and enjoy the conference.

Bremen, June, 15th 2012
Maike Schaefer

Chapter I

Dispersal and hybridisation of GMO

Subspontaneous glyphosate-tolerant genetically engineered *Brassica napus* L. along Swiss railways

Nicola Schoenenberger[1] & Luigi D'Andrea[2]

[1] InnovaBridge Foundation, Caslano; [2] Biome, Delémomt; Switzerland

Extended Abstract[1]

Railway tracks represent a highly interlinked habitat with numerous possibilities for accidental entry of oilseed rape due to seed spill during transportation. Moreover, glyphosate is regularly employed to control the vegetation, increasing the possibility of establishment for plants tolerant to it. We surveyed the presence of genetically engineered glyphosate tolerant oilseed rape (*Brassica napus*) in Swiss railway stations. Our objective was to detect accidental establishment of transgenic plants, since Switzerland does not import nor cultivate transgenic oilseed rape (Swiss Federal Office for Agriculture 2011; Swiss Federal Office of Public Health 2011).

Materials and methods

Leaves from distinct individuals growing in railway stations throughout Switzerland and the Principality of Liechtenstein were analysed using commercially available immunologic test kits, detecting the CP4 EPSPS enzyme, which confers glyphosate tolerance. Positive results were confirmed by a partner organisation by the use of real time PCRs of the *gox* and *cp4epsps* transgenes and an event specific PCR of GT73 glyphosate tolerant oilseed rape.

Results

Seventy-nine railway areas were sampled in Switzerland and the Principality of Liechtenstein, and the feral presence of oilseed rape was detected in 58 of them. A total of 2403 individuals were tested for genetic modification. In four localities, one located in Lugano and three in the area of Basel, a total of 50 plants expressing the CP4 EPSPS protein were detected. In two of the localities, survival of herbicide applications was observed. The populations were probably introduced through contaminated seed spills from freight trains, or during the transfer of goods from cargo ships to trains.

---

1 A full paper is printed to ESEU, Environmental Sciences Europe 2012, 24:23, Series: Implications of GMO-cultivation and monitoring. http://www.enveurope.com/content/24/1/23

Conclusions

Railways are an ideal system for herbicide tolerant GM-plants to establish and spread as a result of high selective pressure in favour of herbicide tolerance with consequent difficulties to keep the infrastructure free of weeds. Crop-to-wild gene flow can occur as several sexually compatible species which are congeneric or in allied genera to oilseed rape were found growing sympatrically (Schoenenberger & Giorgetti-Franscini 2004). Moreover, the capillary presence of railways in the agricultural landscape provides a putative source of contamination of GE-free agriculture. Carefully adapted monitoring designs may be set in order to detect introduction events that can lead to rapid establishment and growing populations as the accepted contamination thresholds are likely to be biologically insufficient to prevent further environmental contamination.

References

Schoenenberger, N., Giorgetti-Franscini, P. (2004) Note floristiche ticinesi: la flora della rete ferroviaria con particolare attenzione alle specie avventizie. Parte II. Boll Soc Tic Sci Nat 92: 97–108.
Swiss Federal Office for Agriculture (2011) Statistik zur Einfuhr von GVO Futtermitteln. http://www.blw.admin.ch
Swiss Federal Office of Public Health (2011) Gesuche und Bewilligungen für GVO-Erzeugnisse. http://www.bag.admin.ch

Variation of maize pollen shedding in North Germany and its relevance for GMO-monitoring

Frieder Hofmann[1,2], Ulrich Schlechtriemen[1,3], Ulrike Kuhn[1,4], Klaus-Peter Wittich[5], Wolfgang Koch[6], Steffi Ober[7], Rudolf Vögel[8] & Mathias Otto[9]

[1] TIEM Integrated Environmental Monitoring, Nörten-Hardenberg/Bremen; [2] Ökologiebüro Hofmann, Bremen; [3] Sachverständigenbüro Schlechtriemen, Nörten-Hardenberg; [4] Büro Kuhn, Bremen; [5] Center of Agro-Meteorological Research, Deutscher Wetterdienst DWD, Braunschweig; [6] Department of Aerosol Technology, Fraunhofer ITEM, Hannover; [7] Nature & Biodiversity Conservation Union NABU, Berlin; [8] Agency for Environment, Health and Consumer Protection, State Brandenburg, Eberswalde; [9] Federal Agency for Nature Conservation (BfN), Bonn; Germany

Abstract

Temporal and spatial variation of maize pollen shedding and dispersal was studied near Angermünde, Brandenburg (GER) in 2010 and 2011. The maize pollen concentration in canopy height was measured with high temporal resolution in and outside of maize fields and comparisons were made between maize fields of different flowering behaviour. The results show a high temporal and spatial variation in pollen concentration. Onset of pollen shedding varied between fields in the same region and same year from mid July to the beginning of August. End of pollen shedding varied from mid to late August. The late-flowering fields tended to release pollen more compactly over 2–3 weeks. In contrast, pollen shedding of the early-flowering fields was interrupted several times due to unfavourable weather conditions, leading to pollen-shedding periods extending to over 5 weeks in the region and over 3 weeks even within the same field.

Introduction

Knowledge of the temporal and spatial variation of maize pollen release rates is a critical component for environmental risk assessment and monitoring of GMO (Aylor et al. 2003; Kawashima et al. 2005). Usually for maize pollen shedding a period of 7–14 days is assumed (EFSA 2009). Nevertheless, there is a lack of precise data mainly due to the fact that direct measurements of pollen release rates in the canopy of maize fields are difficult to perform (Viner et al. 2010).

The Hirst-type volumetric spore and pollen sampler (Hirst 1958) has been the standard device for measurements of pollen concentration in the air for many years. The method has been described in detail e.g. by the British Aerobiology Federation (1995). The device works well outside of fields and is usually operated on higher buildings in cities.

Unfortunately the trap cannot be used for measurements of maize pollen concentration inside a field in canopy height, as the vane movement would be obstructed by the nearby plants. Furthermore, the device is sensitive to turbulent wind conditions which are common in the canopy of maize fields (Shaw et al. 1983; van Hout et al. 2008). Especially greater and heavier pollen like maize pollen are no longer captured in a representative way when the direction of inlet and air movement gets out of line (Hinds 1999; Vincent 2007; Hofmann 2007). To overcome these difficulties a new volumetric sampler, the pollen monitor PMO was used allowing to measure maize pollen concentration in the canopy with high temporal resolution.

Fig. 1: The new pollen monitor PMO, a high-volume sampler with omni-directional inlet for continuous measurement of pollen concentration in the air even under turbulent conditions like inside of maize fields in canopy height or close to the field.

Methods

The maize pollen concentration in the air was measured continuously inside and outside of maize fields near Angermünde, Brandenburg (GER), in two consecutive years (2010 and 2011) using the pollen monitor PMO[1] (Fig. 1). The PMO is a high volume sampler with an omni-directional inlet operated with 1,000 l/min. A bye-pass with 10 l/min leads to an impaction unit, for which the Sporewatch[2] was taken oriented with the inlet (2 mm x 14 mm) in line of the airflow inside the tube. The pollen are deposited on a sticky tape (Silkostrip[3]) placed on a rotating drum. The device is regulated electronically and was operated in a 7-day-modus. Inside the field, the height of the inlet was daily adjusted to the height of the pollen releasing tassels in the canopy. The height of the inlet of the samplers outside of the fields was standardized to 1.8 m. The maize pollen

---
1 TIEM technic GbR, Nörten-Hardenberg (GER)
2 Burkard Scientific Ltd., Uxbridge (GB)
3 Lanzoni s.r.l., Bologna (IT)

deposited on the tape were counted microscopically using the total surface equivalent to an air volume of 14.4 m³/day. The analysis was done in transverse traverses to get hourly data of maize pollen concentration (British Aerobiology Federation 1995).

In 2010 three pollen monitors were placed inside and outside of a 110-ha maize field (A) (Fig. 2a). One pollen monitor was operated inside the field (#1) and two outside in distances of 10 m (#2) and 180 m (#3). In 2011 three fields with differences in their flowering behavior were monitored (Fig. 2b). One PMO (#4) was situated in an early-flowering field (B) and two PMOs were located in comparably late-flowering fields (#5 in C and #6 in D). Furthermore meteorological data were recorded: In 2010 two 3-dimensional ultrasound-anemometers were placed inside field A (site #1) and outside in 50 m distance. In 2011 the weather data of the nearby station of the DWD at Angermünde were used.

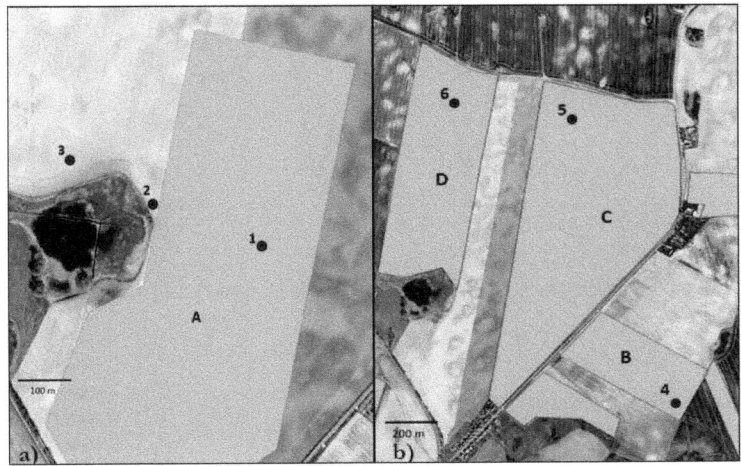

Fig. 2: Measurement sites inside and outside of maize fields near Angermünde, Brandenburg (GER): a) 2010, b) 2011.

Results

Mean daily maize pollen concentrations of field A in 2010 (#1) indicate a compact flowering period starting on August 3rd (Fig. 3) with a main pollen shedding period of 14 days until August 17th, when the pollen release declined to a medium level until the end of measurements on August 22nd. This pattern was reflected by the data of the pollen monitors #2 in 10 m distance and #3 in 180 m distance of the field edge on a respectively lower concentration level. In 2011 pollen measurements in the late-flowering fields C (#5) and D (#6) show a similar pattern of compact pollen shedding lasting from end of July (26th) to mid August (11th) over 2 ½ weeks (Fig. 4). In contrast the early-flowering field B (#4) where the pollen data record started on July 13th reveals a complex pattern of the main pollen shedding period lasting for more than 3 weeks indicating several peaks and interruptions of the pollen release rates due to unfavourable weather conditions. Overall, the results demonstrate that maize pollen shedding may last in a region for more than 5 weeks and even within a single field for over 3 weeks.

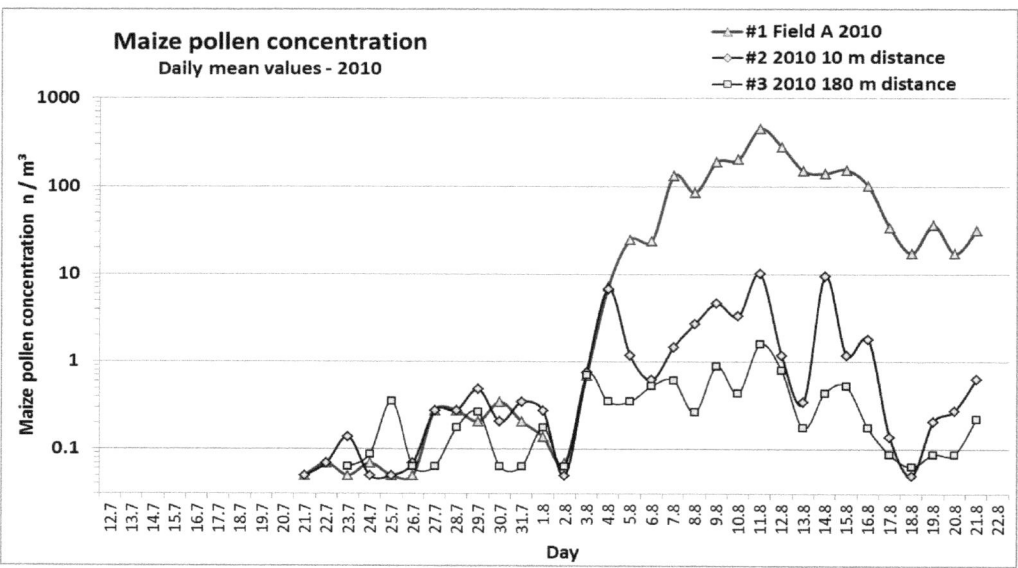

Fig. 3: Daily means of maize pollen concentration inside and outside of a maize field in 2010.

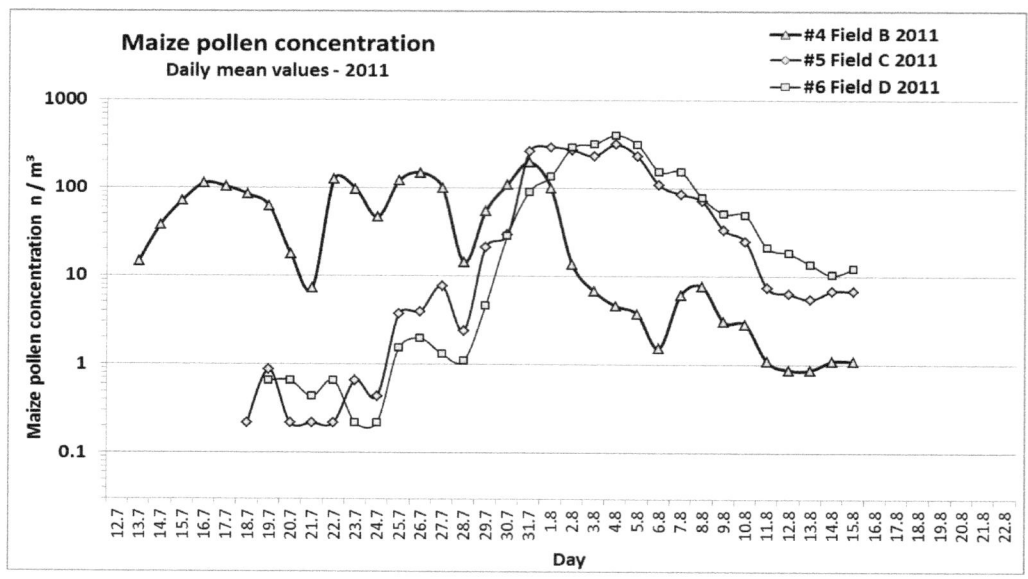

Fig. 4: Daily means of maize pollen concentration inside of maize fields in 2011. B: early-flowering maize field. C+D late-flowering maize fields.

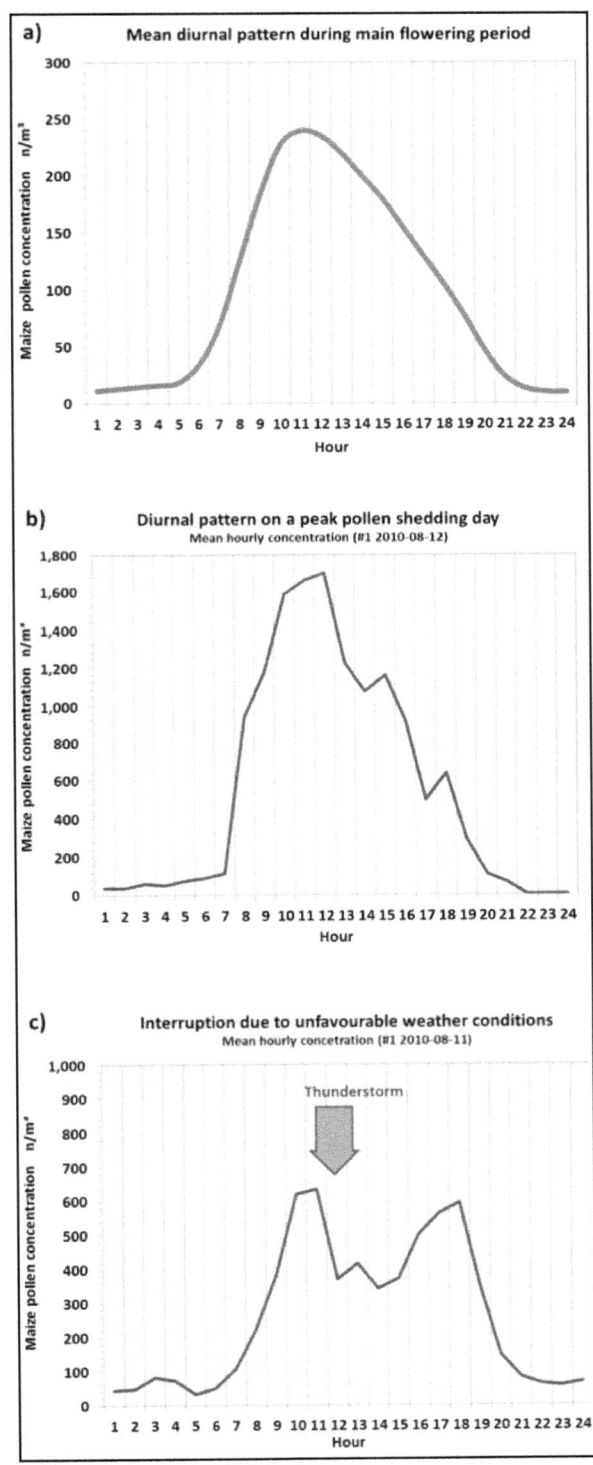

Fig. 5: Diurnal patterns of maize pollen release.

Figure 5a outlines the mean diurnal pattern over the main flowering period as mean hourly concentrations. As described in the literature, the pollen release starts in the early morning and increases rapidly with peak values around midday and declines towards the evening. Maximum pollen concentration on a peak shedding day reached more than 1,700 n/m³ per hour (Fig. 5b). Figure 5c shows the interruption of pollen release due to a thunderstorm. Soon afterward the weather improved and drying winds opened the anthers with a second peak of pollen release in the afternoon. Unfavourable weather conditions could also be identified as the reason for the interruptions in pollen release at the early-flowering field B (#4 in Fig. 4). The analysis of pollen concentration and meteorological data gave no direct correlation to temperature, humidity nor wind speed as single factors, but a positive correlation with turbulence parameters (heat flux density and friction velocity).

Discussion & Conclusions

Our results showed clearly that pollen shedding under the climatic conditions of Northern Germany may last for over 5 weeks in a region and even over 3 weeks within a single field in contrast to a 7–14 day period generally assumed in environmental risk assessment of GMO (EFSA 2009). The detailed data on maize pollen concentration reveals that frequent interruptions of pollen release occur due to unfavourable weather conditions leading to an overall extension of the pollen shedding period. The results are in concordance with earlier reports of long-term measurements of maize pollen concentration in the air at the reference station for rural areas near Ganderkesee, Lower Saxony (GER) over the years 1994 to 2007, where maize pollen could be found frequently in the air over 5–8 weeks (Hofmann et al. 2009). Also Viner et al. (2010) reported records of pollen release rates with more than one peak per day but they could not identify clearly if this was triggered by climatic factors such as temperature. Further analysis of our data on meteorology showed positive correlations of pollen release rates to heat flux density and friction velocity as parameters of turbulence. Turbulence is discussed as a main factor driving pollen release (Boehm et al. 2010).

In respect to environmental risk assessment periods of more than 5 weeks have to be taken into account for maize pollen exposure at least for climatic regions like Northern Germany with temporarily unfavourable weather conditions.

Acknowledgements: This work was funded by the Ministry of Environment, Agriculture and Consumer Protection of Brandenburg, Germany and its Environmental Agency (LUGV), by the Federal Agency of Nature Conservation (BfN), by the Federal Ministry for the Environment, Nature Conservation and Nuclear Safety (BMU). We are grateful for the support of the farmer J. Niedeggen and his team from Gut Kerkow, the NABU Center "Blumberger Mühle", family Winkler from Kerkow, H. Lödding from Fraunhofer ITEM, M. Meyer, A. Meßling, R. Hennings, T. Vogt and H. Salinski from the DWD in Braunschweig and G. Sperling and his team from the DWD weather station Angermünde.

References

Aylor, D.E., Schultes, N.P., Shields, E.J. (2003) An aerobiological framework for assessing cross-pollination in maize. Agricultural and Forest Meteorology 119: 111–129.
Boehm, M.T., Aylor, D.E., Shields, E.J. (2008) Maize Pollen Dispersal under Convective Conditions. Journal of Applied Meteorology and Climatology, 47 (1): 291–307.
EFSA (2009): Scientific opinion of the Panel on Genetically Modified Organisms. The EFSA Journal 1149: 1–85.
Hinds, W.C. (1999) Aerosol Technology. New York, John Wiley & Sons.
Hirst, J.M. (1958) An automatic volumetric spore trap. Annals of Applied Biology 39: 257–265.
Hofmann, F. (2007) Kurzgutachten zur Abschätzung der Maispollendeposition in Relation zur Entfernung von Maispollenquellen mittels technischem Pollensammler PMF. BfN, Bonn. http://www.bfn.de/fileadmin/MDB/documents/themen/agrogentechnik/07-05-31_Gutachten_Pollendeposition_end.pdf (last access 2012/8/15).
Hofmann, F., Janicke, L., Janicke, U., Wachter, R., Kuhn, U. (2009a) Modellrechnung zur Ausbreitung von Maispollen unter Worst-Case-Annahmen mit Vergleich von Freilandmessdaten. BfN Bonn. http://www.bfn.de/fileadmin/MDB/documents/service/Hofmann_et_al_2009_Maispollen_WorstCase_Modell.pdf

Kawashima, S., Matsuo, K., Du, M., Takahashi, Y., Inoue, S., Yonemura, S. (2005) An algorithm for estimating potential deposition of corn pollen for environmental assessment. Environmental Biosafety Research 3: 197–207.

Shaw, R.H., Tavangar, J., Ward, D.P. (1983) Structure of Reynolds stress in a canopy layer. Journal of Climate & Applied Meteorology 22: 1922–1931.

The British Aerobiology Federation (1995) The pollen count guide. National Pollen and Hayfever Bureau, Rotherham UK.

Van Hout, R., Chamecki, M., Brush, G., Katz, J., Parlange, M.B. (2008) The influence of local meteorological conditions on the circadian rhythm of corn (*Zea mays* L.) pollen emission. Agricultural & Forest Meteorology 148: 1078–1092.

Vincent, J.H. (2007) Aerosol sampling: Science, Standards, Instrumentation and Applications. Chichester, John Wiley & Sons.

Viner, B.J., Westgate, M.E., Arritt, R.W. (2010) A model to predict diurnal pollen shed in maize. Crop Science 50: 235–245.

Wind-mediated pollen dispersal of oilseed rape – an estimation using pollen traps

Wieslawa Poplawska[1], Alina Liersch[1], Malgorzata Jedryczka[2], Joanna Kaczmarek[2], Joanna Wolko[1], Maria Ogrodowczyk[1] & Iwona Bartkowiak-Broda[1]

[1] Plant Breeding and Acclimatization Institute, National Research Institute, Poznan;
[2] Institute of Plant Genetics of the Polish Academy of Sciences, Poznan; Poland

Abstract

Oilseed rape (*Brassica napus* L. ssp. *oleifera* Metzg.) is both self-and cross-pollinated. A large number of studies which have measured pollen flow shows a wide variation in oilseed rape pollen dispersal distance and outcrossing rate which have been influenced by different factors such as local climatic conditions (e.g. wind direction, wind speed, temperature, humidity, rainfall), experimental design (e.g. size and orientation of fields) and insect movements. The objective of this study was to assess the distance of the wind-mediated oilseed rape pollen dispersal in Polish environmental conditions using pollen traps and to estimate the isolation distances needed in the case of different types of varieties: conventional, GMO, organic and sowing seed production. The experimental oilseed rape field (acreage about 0,6 ha) was isolated by at least 5 km from other oilseed rape plantations. The pollen traps were placed at five compass directions around the field at the different distances in increments of 5 m. Hirst-type volumetric pollen traps (Burkard Manufacturing, UK) were positioned on a linear transect in the direction of the prevailing wind (S-E) at distances of 90 and 180 m from the edge of the field. Calculation of pollen amount and pollen concentration was based on pollen number visualized on Vaseline-covered and stained Melinex tapes that were mounted on microscope slides and quantified with light microscopic techniques. The obtained results indicate that pollen transfer through wind in oilseed rape was significant for short distance. The highest amount of pollen was generally found in the first 50 m from the edge of field, at each compass direction, a decrease in intensity with increasing distance was observed . However, in the peak of flowering, between 13 and 15 May 2011, the pollen flows up to 180 m of the investigated distance.

Introduction

Oilseed rape (*Brassica napus* L. ssp. *oleifera* Metzg.) is both self-and cross-pollinated. The frequency of outcrossing is dependent on the genotype of the cultivar, environmental growing conditions and the compensatory capacity of the crop (Williams et al. 1987, Mesquida et al. 1988, Free 1993, Westcott & Nelson 2001). Cross pollination is estimated to be between 0 and 90 % depending on the cultivar (Becker et al. 1992; Pierre & Renard 2010). Outcrossing mainly occurs between neighbouring plants in the same field and/or insect pollination and whose pollen can also become airborn and potentially travel at least several kilometres downwind (Treu & Emberlin 2000; Rieger et al. 2002).

The extent of gene flow of genetically different types is largely dependent on the scale of pollen production and dispersal and on the distance between fields in a particular season and place. The matter of gene flow is particularly relevant to oilseed rape because this species is partially allogamous, and produces a huge quantity of pollen grains, 5×10^{12} per ha (Westcott & Nelson 2001). Over a period of approximately 4 to 5 weeks pollen is dispersed both by wind and insects (Williams et al. 1986; Mesquida et al. 1988). the main insect pollinators are honey bees, bumble bees and other insects (Pierre et al. 2003).

The relative contribution of wind and insects to transport and pollination is as yet unresolved (Cresswell & Osborne 2004; Cresswell et al. 2004). The extent of pollen-mediated gene flow in oilseed rape is strongly dependent on climatic conditions (e.g. wind speed and direction) as well as its pollen characteristics. Rapeseed pollen is relatively large (32–33 μm), heavy and sticky, with viability estimates ranging from 1 to 5 days under natural conditions (Treu & Emberlin 2000). Mesquida & Renard (1982) detected the majority rapeseed pollen 32 m from oilseed rape plot, but noted that the concentration of pollen collected decreased rapidly with the distance from the pollen source. Some studies have indicated that viable pollen can be found 1,5 km from the pollen source (Timmons et al. 1995).

In discontinuous pollen-dispersal experiments, cross-hybridization rate was estimated to be 0,0156 % and 0,0038 % at 200 m and 400 m, respectively (Scheffler et al. 1995), whereas in a continuous pollen dispersal experiment, the frequency decreased sharply to 0,02 % at 12 m and was only 0,00033 % at 47 m from the central plot (Scheffler et al. 1993). The objective of this research project was to assess the distance of the wind-mediated oilseed rape pollen dispersal in Polish environmental conditions and to estimate the isolation distances needed in the case of different types of varieties: conventional, GMO, organic and sowing seed production. .

Material and methods

The experimental oilseed rape field (acreage about 0.6 ha) was located in Poznań (N 52025' E 16053') (Great Poland) and was isolated by at least 5 km from other oilseed rape plantations. The double low winter oilseed rape lines were used as the pollen source. The monitoring of intensity and distance of only wind-mediated pollen oilseed rape dispersal was done in the flowering period from 6 to 25 May 2011 using pollen traps. Passive traps were positioned at a height of 1,30 m above ground level and were placed at five compass directions around the field at the different distances of 40, 60, and 90 m in increments of 5 m (Fig. 1).

The passive traps with a sticky area of 12 cm x 5 cm were collected every 24 hours. Hirst-type seven-day volumetric pollen traps T1, T2 (Burkard Manufacturing, UK) were positioned on a linear transect in the direction of the prevailing wind (S-E) at the distance of 90 and 180 m from the edge of the field at a height of 0,60 m above ground level. This apparatus actively sucks air with spores, pollen grains and other small objects, which then stick to the tape covered with a special type of glue.

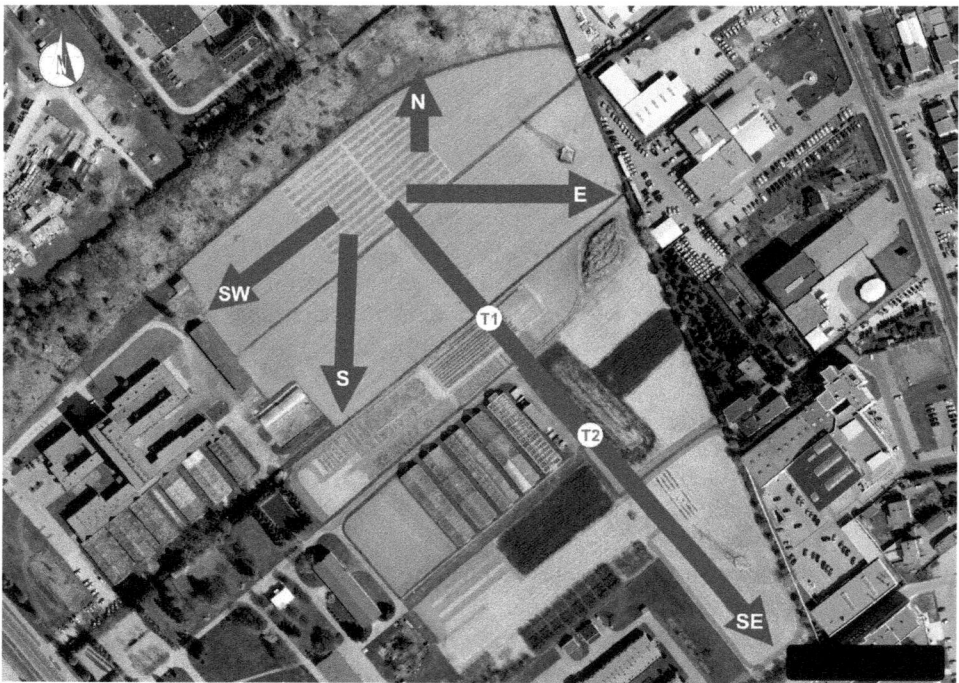

Fig. 1: Layout of the passive and active pollen traps around the experimental field. T-location of a 7-day volumetric pollen sampler (Burkard Manufacturing,UK). T1-distance 90 m from OSR field, T2-distance 180 m from OSR field.

Calculation of pollen amount and concentrations was based on pollen number visualized on Vaseline-covered and stained Melinex tapes that were mounted on microscope slides. The amount of oilseed rape pollen was counted using light microscopy with 250 x magnification. Slides were examined using horizontal counting method – all the oilseed rape pollen grain was counted. Pollen concentration was defined as amount of pollen grain per cubic meter of air sampled, averaged over 24 hours.

Results and discussion

Among factors affecting pollen-mediated gen flow such as environmental conditions, plant species or variety and density, and insect behaviour (Rieger et al. 2002), in this study wind was the only factor for the flow of oilseed rape pollen in both short and long distances. Moreover, only the distance of oilseed rape pollen flow was assessed by using pollen traps and not the outcrossing rate between fields.

A number of other studies has documented pollen wind dispersal in oilseed rape (Treu & Emberlin 2000; Rieger et al. 2002; Hoyle & Cresswell 2007). The dispersal range of pollen varies from a few meters to several kilometres, and the dispersal distances vary with the topographic conditions (Rieger et al. 2002). In general, wind-born pollen flow plays a minor role in long-distance pollination. Most wind-dispersed pollen travels less than 10 m but longer distance transfer can occur under strong winds, and the amount of pollen decreases as the distance from the pollen source increases (Scheffler et al. 1993; Timmons et al. 1995).

Our results showed dispersal events at all distances screened, up to 40 m, 60 m, 90 m (Tab. 1, Fig. 2), and 180 m (Tab. 2). The majority was concentrated in the first few meters around the field of pollen source. The obtained results indicate that pollen transfer through wind in oilseed rape was significant for short distances. The highest amount of pollen was generally found in the first 50 m from the edge of field. In each compass direction, a decrease in intensity with increasing distance was observed (Tab. 1, Fig. 2). However, in the peak of flowering, between 13 and 15 May 2011 the pollen flew up to 180 m of the investigated distances (Tab. 2).

Tab. 1: Amount of oilseed rape pollen grains in the air, captured by using passive traps between 6 and 25 May 2011.

Compass direction	Distance from oilseed rape field (m)																		
	0	5	10	15	20	25	30	35	40	45	50	55	60	65	70	75	80	85	90
N	1272	767	601	541	304	210	103	54	79	0	0	0	0	0	0	0	0	0	0
E	895	849	252	299	194	242	151	52	117	25	79	86	58	30	119	90	20	5	8
S	490	318	143	120	93	125	10	46	19	87	46	62	96	1	76	20	14	25	0
SW	271	339	115	284	146	42	44	45	20	0	14	15	2	0	0	0	0	0	0
SE	24	167	234	60	103	85	108	89	68	71	28	62	31	50	68	47	38	79	9

The pollen amounts show a typical leptokurtic curve in function of distance. Most dispersal functions are "leptokurtic", with dispersal to both short and long distances occurring more frequently than they would do under a Gaussian function (Clark, Lewis & Horvath 2001). Pollen flow can be used as an indicator as it gives the potential for cross pollination and fertilization to occur. However, several other factors are important for cross-fertilization to occur (e.g. species must be sexually compatible, flowering time needs to overlap, pollen must be viable when landing on receptor plant's stigmas and must be able to compete with self-pollen).

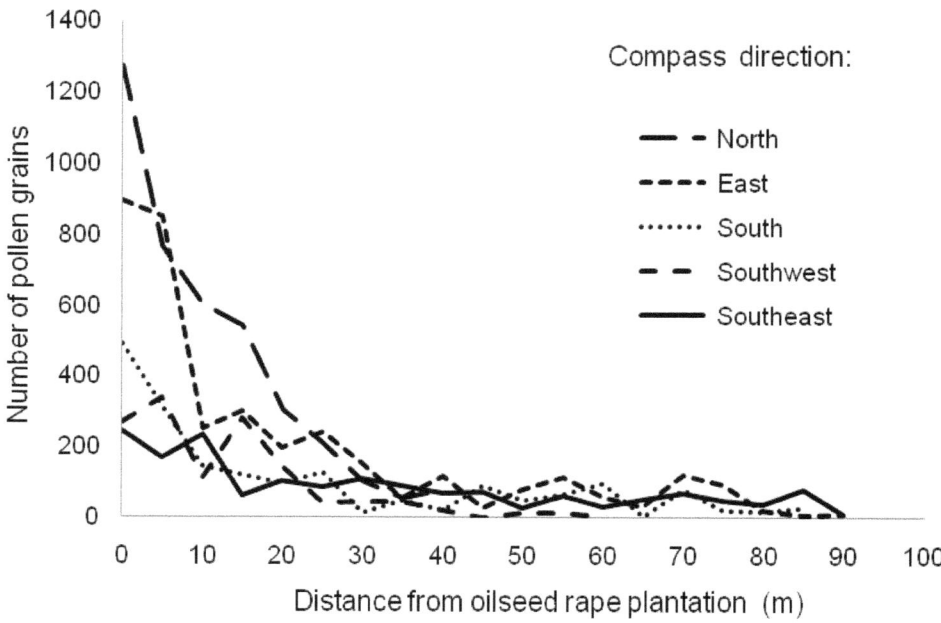

Fig. 2: Amount of oilseed rape pollen grains at different distances from the OSR field, captured by using passive traps between 6 and 25 May 2011.

A number of studies have reported maximum distance of up to 1.5 km for wind-mediated pollen dispersal by measuring directly for oilseed rape pollen (Timmons et al. 1995) and between 400 m and 4 km for pollination by using male sterile or emasculated bait plants to detect pollination (Timmons et al. 1995; Thompson et al. 1995; Norris et al. 1999). Studies using male sterile bait plants, only represent the potential for gene flow, with pollination levels for male fertile plants likely to be much lower.

The distance and success to which pollen mediated gene flow is likely to occur is dependent not only on its dispersal in space, by either wind or insect action, but also on the length of time the pollen grain retains its potential for pollination. Pollen viability varies with environmental conditions, particularly temperature and humidity.

Under controlled conditions in the laboratory, oilseed rape pollen can remain viable for between 24 hours up to one week (Mesquida & Renard 1982). Under natural conditions pollen viability gradually decreases over 4 to 5 days. The flowers of oilseed rape produce nectar with relatively high concentrations of sugars and have a colour and structure which makes them attractive to insects, particularly honeybees. Many studies have shown that a large proportion (up to 80 %) of bees flights are less than 5 m (Pierre et al. 2003; Cresswell & Osborne 2004). Occasionally however, bees may travel much further with distances of 1 to 2 km, up to a maximum distance of 4 km (Ramsay et al. 1999; Thompson et al. 1999).

A large number of studies which have measured pollen flow shows a wide variation in oilseed rape pollen dispersal distance and outcrossing rate which have been influenced by different factors such as local climatic conditions (e.g. wind direction, wind speed, temperature, humidity, rainfall), experimental design (e.g. size and orientation of fields) and insect movements (Scheffler et al. 1993).

Tab. 2: Concentration of OSR pollen grain in the air, observed by using volumetric traps.

	Distance of Hirst-type volumetric pollen trap from the field of oilseed rape			
	90 m		180 m	
Date of observation	Number of pollen grains observed in the microscope slide	Number of pollen grains per 1 m^3 of air	Number of pollen grains observed in the microscope slide	Number of pollen grains per 1m^3 of air
13.05.2011	92	12.78	11	1.53
14.05.2011	6	0.83	2	0.28
15.05.2011	11	1.53	1	0.14
16.05.2011	6	0.83	0	0.00
17.05.2011	0	0.00	0	0.00
18.05.2011	3	0.42	0	0.00
19.05.2011	0	0.00	0	0.00
20.05.2011	12	1.67	0	0.00
21.05.2011	0	0.00	0	0.00
22.05.2011	0	0.00	0	0.00
23.05.2011	0	0.00	0	0.00
24.05.2011	0	0.00	0	0.00

Conclusions

The obtained results indicated that pollen wind-dispersal of oilseed rape is limited to relatively short distance, therefore cross pollination which is the result of the transfer of pollen by wind is not the main problem for coexistence of different types of oilseed rape cultivars. The transfer of different genotypes by pollen dispersal can be managed through spatial separation and the use of buffer or discard zones where crops are in close proximity.

Acknowledgements

This project was supported by the Polish Ministry of Science and Higher Education, project no. PBZ 06/1/2007

References

Becker, H.C., Karle, R., Han, S.S. (1992) Environmental variation for outcrossing rates in rapeseed (*Brassica napus*). Theoretical and Applied Genetics. 84: 303–306.
Clark, J.S., Lewis, M., Horvath, L. (2001) Invasion by extremes: population spread with variation in dispersal and reproduction. American Naturalist. 157: 537–554.
Cresswell, J.E, Osborne, J.L. (2004) The effect of patch size and separation on bumblebee foraging in oilseed rape: implications for gene flow. Journal of Applied Ecology. 41: 539–546.
Cresswell, J.E, Davies, T.W., Patrick, M.A., Russell, F., Pennel, C., Vicot, M., Lahoubi, M. (2004) Aerodynamics of wind pollination in a zoophilous flower, *Brassica napus*. Functional Ecology. 18: 861–866.
Free J.B. (1993) Insect pollination of crops. Academic press, London, UK.
Hoyle, M., Cresswell, J.E. (2007) The effect of wind direction on cross-pollination in wind-pollinated GM crops. Ecological Applications. 17: 1234–1243.
Mesquida, J., Renard, M. (1982) Étude de la dispersion du pollen par le vent et de l'importance de la pollinisation anémophile chez le colza (*Brassica napus* ssp. *oleifera* Metzger). Apidologie. 13: 353–366.
Mesquida, J., Renard, M., Pierre, J.S. (1988) Rapeseed (*Brassica napus* L.) productivity; the effect of honeybees (*Apis mellifra* L.) and different pollination conditions in cage and field tests. Apidologie. 19: 51–72.
Norris, C.S., Simpson, E.C., Sweet, J.B., Thomas, J. E., (1999) Monitoring weediness and persistence of genetically modified oilseed rape. In Gen Flow and Agriculture: Relevance for Transgenic Crops. University of Keele, Staffordshire: 255–260.
Pierre, J., Renard, M. (2010) Bilan de 30 ans de travaux de recherché effectues en France sur la pollinisation du colza. OCL. Vol.17 No 3: 121–125.
Pierre, J., Marsault, D., Genecque, E., Renard, M., Champolivier, J., Pham-Delegue, M.H. (2003) Effects of herbicide-tolerant transgenic oilseed rape genotypes on honey bees and other pollinating insects under field conditions. Entomologia Experimentalis et Applicata. 108: 159–168.
Ramsay, G., Thompson, C.E., Neilson, S., Mackay, G.R. (1999) Honeybees as vectors of GM oiseed rape pollen. Monograph-British Crop Protection Council. 72: 209–214.
Rieger, M.A., Lamond, M., Preston, C., Powles, S.B., Rousch R.T. (2002) Pollen-mediated movement of herbicide resistance between commercial canola fields. Science. 296: 2386–2388.
Scheffler, J.A., Parkinson, R., Dale, P., (1993) Frequency and distribution of pollen dispersal from transgenic oilseed rape (*Brassica napus*). Transgenic Research. 2: 356–364.
Scheffler, J.A., Parkinson, R., Dale, P. (1995) Evaluating the effectiveness of isolation distance for field plot of oilseed rape (*Brassica napus*) using a herbicide-resistance transgene as a selectable marker. Plant Breeding. 114: 317–321.

Timmons, A.M., O'Brajen, E.T., Charters, Y.M., Dubbels, S.J., Wilkinson, M.J. (1995) Assessing the risks of wind pollination from fields of genetically modified *Brassica napus* ssp. *oleifera*. Euphytica. 85: 417–423.

Thompson, C.E., Squire, G., Mackay, G.R., Bradshaw, J.E., Crawford, J., Ramsay, G. (1999) Regional patterns of gene flow and its consequences for GM oilseed rape. Gene Flow and Agriculture: Relevance for Transgenic Crops, BCPC Symposium Proceedings No. 72, Keele, UK, pp. 95–100.

Treu, R., Emberlin, J. (2000) Pollen dispersal in the crops Maize (*Zea mays*), Oilseed rape (*Brassica napus* ssp. *oleifera*), Potatoes (*Solanum tuberosum*), Suger beet (*Beta vulgaris* ssp.*vulgaris*) and Wheat (*Triticum aestivum*). A Report for the Soil Association from the National Pollen Research Unit.

Williams, I.H., Martin, A.P., White, R.P. (1986) The pollination requirements of oil-seed rape (*Brassica napus* L.). Journal of Agricultural Science. 106: 27–20.

Williams, I.H., Martin, A.P., White, R.P. (1987) The effect of insect pollination on plant development and seed production in winter oil-seed rape (*Brassica napus* L.). Journal of Agricultural Science. 109: 135–139.

Westcott, L., Nelson, D. (2001) Canola pollination: an update. Bee World. 82: 115–129.

Breckling, B. & Verhoeven, R. (2013) GM-Crop Cultivation – Ecological Effects on a Landscape Scale.
Theorie in der Ökologie 17. Frankfurt, Peter Lang.

Investigation of oilseed rape gene flow using erucic acid as biochemical marker

Iwona Bartkowiak-Broda, Wieslawa Poplawska, Alina Liersch, Tadeusz Walkowski & Maria Ogrodowczyk

Plant Breeding and Acclimatization Institute, National Research Institute, Poznan, Poland

Introduction

Oilseed rape (*Brassica napus* L. ssp. *oleifera* Metzg.) has become an important crop over the past 30 years. At present, it is second (13.9 %) in the world regarding the production of oil seeds, following soybeans (60.3 %), and third (15.3 %) regarding the production of vegetable oil, following palm oil (34.2 %) and soybean oil (27.3 %) (Oil World 2010). The market demand of rapeseed oil is continually increasing for nutritional purposes and for different technologies, especially for biodiesel production.

Due to its economical importance, rapeseed is one of the principal GM crops, besides soybean, cotton and corn. In 1995, first GM cultivars were registered: Quest (Monsanto), resistant to glyphosate, and Innovator (Aventis), resistant to glufosinate ammonium. Since 1995, more than 100 GM cultivars have been registered being tolerant to nonselective herbicides like glufosinate (commercial name Basta, Liberty), glyphosate (Roundup), and bromoxynil (no longer available), or expressing modifications of fatty acid patterns, and a hybridization system based on GM mediated male sterility. Regarding GM oilseed rape, new traits were developed, for instances, producing low levels of saturated fatty acids, high levels of short and medium unsaturated fatty acids, novel fatty acids, a resistance to drought and insects, and an increase of bioactive compounds in oil (Friedt & Lühs 1999).

The cultivation of genetically modified herbicide resistant oilseed rape has increased over the past few years, especially in North America, Asia and Australia. The acreage of GM oilseed rape in the world amounts 8.2 Mio ha, which was 5 % of the total acreage of GM crops in the world (ISAAA 2011).

In Europe, GMOs cannot be put on the market without an approval. The essential principles of the EU policies are strict safety standards and safeguarding of freedom of choice for consumers and for farmers. However, GM oilseed rape is considered of being not suitable for coexistence. Important obstacles concerning the integration of GM cultivars into the cropping system in Europe are: the flow of transgenic pollen to the surroundings and potential cross pollination; transfer by seed dropped at harvest and during transport and the capacity of seed persistence in soil which may lead to volunteers over many years. However, taking into account the quick development of breeding of GM and non-GM oilseed rape cultivars of different traits, the problems of coexistence of both types of cultivars are also investigated in Polish agricultural and environmental conditions.

Methods

The investigations concerning winter oilseed rape volunteers have been conducted in two seasons in three voivodeships in Poland situated along the coast of the Baltic Sea. The research was conducted in both seasons on 27 plantations of double low winter oilseed rape cultivars (Castille, Californium, Lisek, Rasmus) at 17 farms in 2004/2005 and at 21 farms in 2005/2006. The harvested seeds of double low cultivars cannot contain more than 2% of erucic acid and the glucosinolate content is limited up to 25 $\mu M\ g^{-1}$ free fatty dry matter in seeds. The development of over 2,000 plants of rapeseed was observed showing both typical and abnormal growth morphology. Before flowering, the plants were isolated with bags and seeds were harvested after the maturity in both seasons from 1,588 plants.

The erucic acid was used as biochemical marker of gene flow by pollen and seeds. Erucic acid content in rapeseed's oil is determined by embryo genes (two pairs of alleles with additive effect) and is not influenced by environmental conditions (Harvey & Downey 1964). Glucosinolate content was taken into account as an additional trait. Erucic acid content was measured using the method of fatty acid methyl esters (Byczynska & Krzymanski 1969), and glucosinolates were measured using the method of silyl derivatives content (Michalski, Kołodziej & Krzymanski 1995), both by gas chromatography. For the estimation of plants ploidy, cytometric analyses of relative nuclear DNA content were performed.

Results

The variability of the quality of investigated plants was very high as it is shown in Table 1 for selected plant material divided into two groups: plants with morphology similar to rapeseed and plants similar to turnip rape.

In the group of rapeseed like plants seeds characterized by low and very high erucic acid content, up to 49.6%, were found. It indicates that these seeds originate from old high erucic varieties and from hybrids of plants with genome determining high- or low erucic acid content in seeds. Similarly, the broad range of glucosinolates content in seeds confirms the origin of investigated plants. Also some plants are the result of crosses between rapeseed and turnip rape what indicates the nuclei DNA content ranging from 46.1 (fluorescence channel number typical to turnip rape) to 128.0 – typical to rapeseed.

Table 1. Erucic acid and glucosinolates content and relative DNA nuclei content in seeds of winter oilseed rape volunteers. 00: double low rapeseed; 0HG: low erucic, high glucosinolate rapeseed; HE0: high erucic, low glucosinolate rapeseed; HEHG: high erucic, high glucosinolate rapeseed. (acc. to Popławska & Bartkowiak-Broda 2004; Liersch et al. 2008)

Items	Number of populations	Erucic acid [%]		Glucosinolates [$\mu mol \cdot g^{-1}$ seeds]		Relative DNA content of nuclei - fluorescence channel number	
		mean	range	mean	range	mean	range
Rapeseed – like plants							
Rapeseed "00"	2	0.0	0	19.6	13.4–25.8	119.0	114.6–123.4
Rapeseed "0HG"	37	0.08	0–1.8	73.0	39.4–108.8	117.7	112.0–128.0
Rapeseed "HE0"	6	24.5	2.7–45.5	12.7	10.3–16.5	117.9	115.2–123.1
Rapeseed "HEHG"	48	25.7	7.6–49.6	72.0	26.8–102.8	116.4	51.2–128.0
cv. Castille		0.0	–	12.4		116.9	
cv. Californium		0.0	–	14.8		118.4	
cv. Lisek		0.0	–	9.9		115.6	
cv. Rasmus		0.0	–	9.9		114.3	
Turnip rape like-plants	29	39.9	26.6–47.8	90.9	42.5–135.4	50.8	46.1–119.1
Turnip rape cultivars:							
cv. Brachina		39.5	–	44.3		51.0	
cv. Ludowy		39.5	–	44.3		49.4	

On the basis of the obtained results it was possible to state that the investigated plants with higher level of erucic acid and glucosinolates in comparison to double low cultivars are:
- volunteers of old cultivars of high erucic and high glucosinolates,
- hybrids among volunteers of old and double low (low erucic and low glucosinolate content) cultivars,
- volunteers of relatives from Brassica genus, especially volunteers of turnip rape,
- recombinants between oilseed rape and turnip rape.

Conclusion

Seeds of rapeseed can survive in soil for many years. In the soil seed bank the seeds of old varieties are still present despite the fact that no erucic cultivars were sown since 1985 and since 1990 only double low cultivars were cultivated in Poland.

The obtained results, using of erucic acid as biochemical marker, confirm the results of the investigations of other authors that one of the most important problem for coexistence is due to volunteers and feral populations linked to the phenomenon of seed secondary dormancy.

Acknowledgements: This project was partly supported by the Ministry of Science and Higher Education, project no. PBZ 06/1/2007 and 6 FP EU SIGMEA

References

Byczynska, B., Krzymanski, J. (1969) A fast method for obtaining of fatty acids methyl esters to be analysed by gas chromatography method (in Polish). Tluszcze Jadalne XIII: 108–114.

Friedt, W.F. & Lühs, W.W. (1999) Breeding of rapeseed (Brassica napus) for modified seed quality – synergy of conventional and modern approaches. Proc. of 10th Rapeseed Congress 26–29 September, Canberra, Australia, CD.

Harvey, B.L., Downey, R.K. (1964) The inheritance of erucic acid content in rapeseed (*Brassica napus*). Canadian Journal of Plant Science 44: 104–111.

Michalski, K., Kołodziej, K., Krzymanski, J. (1995) Quantitative analysis of glucosinolates in seeds of oilseed rape. Effect of sample preparation on analytical results. Proc. of 9th Intern. Rapeseed Congress, Cambridge, UK, 3: 911–913.

Oil World (2010) www.worldoil.com. 17/2010

Poplawska, W., Bartkowiak-Broda, I. (2004) Reasons of quality decrease of oil row material from seeds of rapeseeds (in Polish). Rosliny Oleiste – Oilseed Crops XXV/2: 495–503.

Liersch, A., Poplawska, W., Ogrodowczyk, M., Bartkowiak-Broda, I., Bocianowski, J. (2008) Phenotypic acharacteristics of winter oilseed rape volunteers (*Brassica napus* L.) occurring in northern regions of Poland (in Polish). Biuletyn IHAR: 250: 249–260.

The influence of volunteers and soil seed bank on the quality of oilseed rape seeds

Alina Liersch, Joanna Wolko, Wieslawa Poplawska, Krystyna Krotka & Iwona Bartkowiak-Broda

Plant Breeding and Acclimatization Institute, National Research Institute, Poznan, Poland

Abstract

Oilseed rape (*Brassica napus*, OSR) displays a high frequency of seed shattering before and during harvest. Seed shattering is more than an economic loss because seeds can fall dormant, survive for many years in soil and emerge in the following crops. The objective of this study was the observation of volunteers of winter OSR derived from soil seed bank in the field after the harvest of high erucic acid (HEAR about 57%) cultivar Maplus in two localities: Dlon, Zielecin. In the seeds of 66 volunteers derived from the soil seed bank erucic acid and glucosinolates (GLS) content have been analysed. Genomic DNA of all volunteers was analysed using the molecular marker system RAPD (25 starters). The results showed that the volunteers obtained from the soil seed bank were composed of varieties cultivated in these fields many years earlier – old traditional varieties, variety Maplus, double low varieties and seeds of hybrids between traditional and double low winter OSR varieties.

Introduction

Oilseed rape (OSR) (*Brassica napus* L. ssp. *oleifera* Metzg.) shows a high frequency of seed shattering before and during harvest. The amount of seeds spilled to the soil is between 5% and 10% of the total seed set (typically between 1,000 and 6,000 seeds m^{-2}) (Gruber et al. 2011). At unfavourable weather conditions the spillage can be even more severe. It is a serious economic disadvantage concerning the level and quality of seed yield. These seeds incorporated in the soil seed bank often persist there for a long time, and they are often non released from dormancy before more than 10 years (Lutman et al. 2005). They are the source of volunteers contaminating the harvest of the subsequent oilseed rape plantations with seeds as well as with pollen, which can pollinate the plants of current crop (Bartkowiak-Broda et al. 2008). Moreover, volunteers competing with plants of current plantation for light, water and nutrients, are also the vehicle for pests and diseases for this crop. This is even more likely as crop rotations become shorter (Gruber et al. 2012). Volunteers reduce the potential of seeds yield, which can cause financial losses for farmers but also changes the quality of harvested seeds.

According to investigations conducted in Poland, the influence of volunteers of old varieties can be very high on the quality of seeds. Varieties of high erucic acid (HEAR) (~ 50% of all fatty acids) and high glucosinolate (~ 160 µmol·g^{-1}seeds) were replaced in Poland about 20 years ago by double low winter OSR varieties (erucic acid < 2.0% and glucosinolates content < 15 µmol·g^{-1}seeds). Studies carried out in Poland showed that until now on the same fields, plants and seeds with increased erucic and glucosinolates values occurred. It was stated that the most negative influence on yield quality in case of the coexistence of varieties of different quality traits may affect volunteers with HEAR content or high glucosinolate content (or both) as well as hybrids between traditional and presently cultivated double low varieties (Bartkowiak-Broda et al. 2008). A similar problem of admixtures of different genotypes in the harvested yield can occur in case of coexistence of oilseed rape cultivars with different quality traits (e.g. cultivars with different fatty acid content or yellow seeded and black seeded varieties) as well as genetically modified (GM) and non-GM varieties (Colbach 2011).

The aim of this study was the estimation of soil seed bank abundance – quantity and quality of seeds in the soil seed bank after the harvest of high erucic acid (about 57%) variety Maplus, in different types of soil.

Material and methods

The objects of this study were two experimental field trials located in different environment conditions, Dlon (51°46′N, 17°14′E) and Zielęcin (52°10′N, 16°22′E) and various soil complexes: Dlon – typical heavy soil of quality class III/IVa and Zielecin – light soil of quality class IVa. The acreage of a field trial in each locality was about 9 ha; 1 ha of high erucic acid winter OSR variety Maplus (57% erucic acid and glucosinolates content 12.6 µmol·g^{-1}seeds) was surrounded by 8 ha of double low winter OSR variety Monolit (0% erucic acid and glucosinolates 14.56 µmol·g seeds) (Fig. 1).

After the harvest of the trials from 1 ha of winter OSR crop variety Maplus, 32 round soil cores with 2,5 cm diameter and ca 20 cm depth were taken in Dlon and Zielecin to assess the number and quality of seeds in oilseed rape soil seed bank. The volunteer plants obtained from soil samples were placed in the greenhouse. Before flowering, the plants were isolated with bags. The seeds were harvested after maturity and analysed with biochemical and molecular methods. Seed samples were analysed in respect to erucic acid content using the method of methyl esters of fatty acids (Byczyńska & Krzymański 1969) and glucosinolates (GLS) using the method of gas chromatography of silyl derivatives of desulfoglucosinolates (Michalski et al. 1995). All individual samples of seeds were described using the molecular marker system RAPD.

DNA was extracted from oilseed rape leaves using a modified CTAB procedure according to Doyle and Doyle (1990). The basic RAPD reactions were performed as described by Williams et al. (1990). DNA profiles of the volunteers were compared with DNA profiles on HEAR Maplus and double low winter OSR cultivar Monolit. Genetic distance (GD) of investigated volunteers was calculated according to the Nei and Li (1979) coefficient. The investigated genotypes were grouped using the unweighted pair group method employing arithmetic average (UPGMA). The distances among volunteers were visualized with dendrograms. All calculations were carried out using PHYLIP 3.5 (Felsenstein 1993).

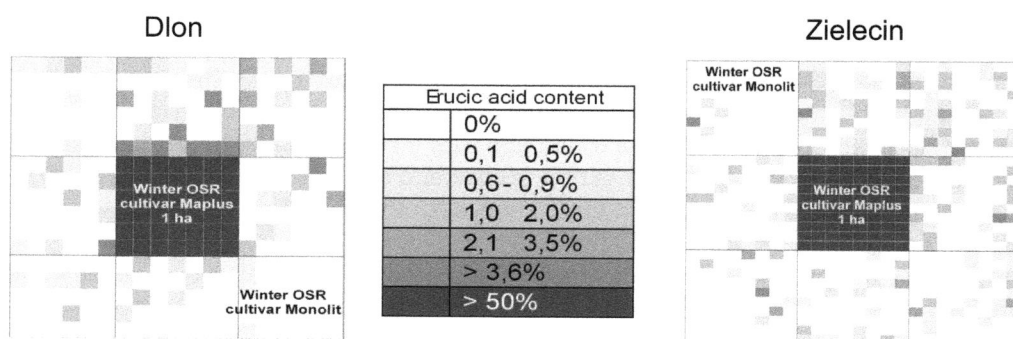

Fig. 1: Erucic acid content of winter OSR from field trials at Dlon and Zielecin

Results

The seeds of 66 volunteer plants (51 volunteers from Dlon: GM75-GM124, DHER and 15 volunteers from Zielecin: GM62-GM74, 126, 127) derived from the soil seed bank were characterized by erucic acid content within the range from 0.0 to 57.4% of all fatty acids and glucosinolates content 7.8 to 105.5 $\mu mol \cdot g^{-1}$ seeds (Tab. 1).

RAPD molecular genetic studies divided seed samples into four major clusters (Fig. 2). Cluster A was subdivided and included different kinds of volunteers: double low OSR, volunteers with high erucic acid content and low glucosinolates content and volunteers originating from hybrids between these cultivars with an erucic acid content from 0.0 to 27.6% of all fatty acids and low glucosinolates (GLS) content (12.6 – 19.0 $\mu mol \cdot g^{-1}$ seeds). Cluster B represents volunteers with high erucic (mean 40.8%) and high glucosinolates content (mean 40.18 $\mu mol \cdot g^{-1}$ seeds). Cluster C and D included the high erucic and low glucosinolate volunteers (52.79% of all fatty acids and 23.67 $\mu mol \cdot g^{-1}$ seeds of glucosinolates) similar to variety Maplus (erucic acid content 52.50% of all fatty acids and glucosinolates content 16.88 $\mu mol \cdot g^{-1}$ seeds).

The volunteers from the soil seed bank indicated the presence of seeds belonging to the variety Maplus, old traditional varieties cultivated many years ego, double low varieties and seeds originating from hybrids between traditional and double low winter OSR cultivars. Especially interesting is the presence of seeds with high erucic and high glucosinolate content despite this type of varieties had not been grown in these fields in the previous 20 years.

Tab.1: Erucic acid and glucosinolates content in seeds from soil seeds bank at Dlon and Zielecin

Sample	Erucic acid [%]	Total GLS [µmol· g⁻¹ seeds]	Sample	Erucic acid [%]	Total GLS [µmol· g⁻¹ seeds]
Cluster A	**7,78**	**16.78**	GM117	55.6	18.2
GM62	2.0	18.6	GM118	54.5	28.1
GM70	3.8	19.0	GM114	54.9	45.6
GM73	2.4	18.4	GM72	54.7	15.8
GM96	27.6	15.7	GM119	52.1	12.5
MO/MAP	26.4	16.5	**Cluster D**	**52.50**	**16.88**
MO.126.127	0.0	12.6	GM64	54.3	15.9
Cluster B	**40.8**	**40.18**	GM77	53.6	18.7
GM66	48.6	105.5	GM116	55.8	20.1
GM67	53.2	93.7	GM79	51.6	17.2
GM69	36.7	87.7	GM120	47.9	20.0
DHER	55.6	5.3	GM105	54.2	16.1
GM95	0.0	13.3	GM107	54.2	13.3
GM102	52.4	28.5	GM115	54.4	18.9
GM111	36.2	23.5	GM74	56.0	17.6
GM121	0.0	85.0	GM90	55.7	13.4
GM125	55.6	5.3	GM103	49.9	24.3
GM78	39.5	12.7	GM75	44.4	23.8
GM97	49.6	86.5	GM76	54.5	23.4
GM105	54.2	16.1	GM93	46.5	12.1
GM108	53.6	16.0	GM100	50.6	23.3
GM109	15.4	15.7	GM92	57.4	16.8
GM110	53.6	16.0	GM94	56.3	15.1
GM112	55.7	13.7	GM123	55.9	14.1
GM113	48.5	13.1	GM124	55.4	20.3
Cluster C	**52.79**	**23.67**	GM80	51.5	17.3
GM63	55.5	27.3	GM88	50.7	17.5
GM65	55.4	26.8	GM81	52.6	9.5
GM71	53.9	22.8	GM82	51.1	7.8
GM68	51.8	23.3	GM84	48.9	17.2
GM122	42.9	18.4	GM87	48.9	12.7
GM104	50.4	17.7	GM89	49.1	11.8
GM91	50.5	15.1	GM83	55.0	14.4
GM99	53.2	17.2	GM85	52.3	16.4
GM101	52.8	48.9	GM86	51.2	19.1
MONOLIT	0.0	14.5	**MAPLUS**	57.0	12.6

These results confirm that the seeds of different quality persist for a long time in the soil seed bank at the stage of secondary dormancy and can be the source of volunteers polluting the yield of oilseed rape. Also, the types of soil and soil environmental conditions specific to soil texture may influence the seed persistence and abundance in the soil seed bank. The number of volunteers derived from the soil seed bank was much higher at Dlon having more heavy soils than at Zielecin, which characterized by a light soil (Tab. 1). This work and other investigations have also suggested that seeds persist for a shorter time in light soils in comparison to heavy soils (Lutman et al. 2002).

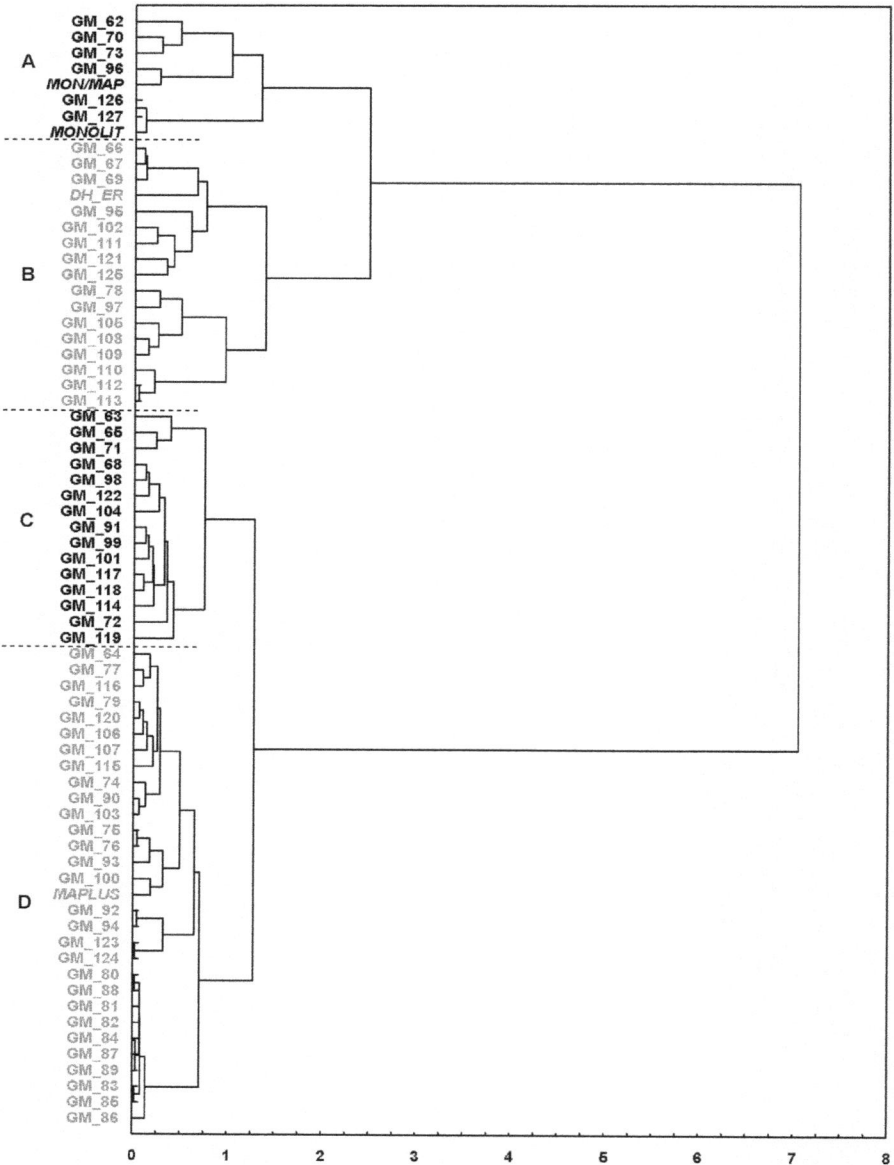

Fig. 2: Dendrogram of 66 volunteers winter OSR from soil seeds bank based on RAPD markers.

Conclusions

The investigations of volunteers and soil seed bank confirmed that spilled seeds and hybrids of volunteers with cultivated varieties can contaminate the yield of seeds. This is the most important problem concerning the coexistence of different quality oilseed rape cultivars as well as GM and non-GM varieties.

Acknowledgements: This project was supported by the Polish Ministry of Science and Higher Education, project no. PBZ 06/1/2007

References

Bartkowiak-Broda, I., Walkowski, T., Poplawska, W., Ogrodowczyk, M., Liersch, A. (2008) Influence of volunteers and oilseed rape-like plants on the quality of winter oilseed rape yield (in Polish). Oilseed Crops, XXIX: 185–196.

Byczynska, B., Krzymanski, J. (1969) A fast method for obtaining of fatty acids methyl esters to be analysed by gas chromatography method (in Polish). Tłuszcze Jadalne, XIII: 108-114.

Colbach, N. (2011) Evaluation of prospective cropping system scenarios for managing oilseed rape volunteers and harvest purity using the GENESYS model. Proc. 13th International Rapeseed Congress, June 5-9, 2011 Praque, Czech Republic, Abstract book: 6–9.

Doyle, J.J. and Doyle, J.L. (1990) Isolation of plant DNA from fresh tissue. Focus 12: 13–15.

Felsenstein, J. (1993) PHYLIP (Phylogeny Inference Package) version 3.5c. Distributed by the author. Department of Genetics, University of Washington, Seattle.

Gruber, S., Feike, T., Weber, E.A., Claupein, W. (2011) Following the trace of harvesting seed losses - a model to predict the soil seed bank and oilseed rape volunteers. Proc. 13th International Rapeseed Congress, June 5-9, 2011 Praque, Czech Republic, Abstract book: 168–171.

Gruber, S., Hüsken, A., Dietz-Pfeilstetter, A., Möllers, C., Weber, E.A., Stockmann, F., Thöle, H., Schatzki, J., Dowideit, K., Renard, M., Becker, H.C., Schiemann, J., Claupein, W. (2012) Biological confinement strategies for seed-and pollen-mediated gene flow of GM Canola (*Brassica napus* L.). AgBioForum. 15(1): 44–53.

Lutman, P.J.W., Berry, K., Payne, R.W., Simpson, E., Sweet, J.B., Champion, G.T., May, M.J., Wightman, P., Walker, K., Lainsbury, M. (2005) Persistence of seeds from crops of conventional and herbicide tolerant oilseed rape (*Brassica napus*). Proceedings of the Royal Society London. B 272, 1909–1915.

Lutman, P.J.W., Cussans, G.W., Wright, K.J., Wilson, B.J., Mcn Wright, G., Lawson, H.M. (2002) The persistence of seeds of sixteen weed species over six years in two arable fields. Weed Res. 42: 231–241.

Michalski, K., Kolodziej, K. and Krzymanski, J. (1995) Quantitative analysis of glucosinolates in seeds of oilseed rape – effect of sample preparation on analytical results. Proc. 9th International Rapeseed Congress, 4-7 July 1995, Cambridge, UK 3: 911–913.

Nei, M., Li, W. (1979) Mathematical model for studying genetic variation in terms of restriction endonucleases. Proceedings of the National Academy of Sciences. USA 76: 5269–5373.

Williams, J.G.K., Kubelik, A.R., Livak, K.J., Rafalski, J.A. and Tingey, S.V. (1990) DNA polymorphisms amplified by arbitrary primers are useful as genetic markers. Nucleic Acids Research. 18: 6531–6535.

Pollen flow evaluation to implement a Quick Monitoring Index (QMI)

Elena Balducci[1], Donatella Paffetti[2], Davide Travaglini[2], Stefano Biricolti[3], Francesca Bottalico[2], Silvia Fiorentini[2], Anna Buonamici[1], Francesca Donnarumma[1], Alessandro Materassi[4], Gianni Fasano[4], Lorenzo Chelazzi[5], Filippo Cimò[1], Isabella Colombini[5], Laura Bartalucci[6], Antonio Perfetti[7], Olga Mastroianni[7], Valeria Tomaselli[8], Simone Gorelli[9], Francesco Tonazzini[9] & Cristina Vettori[1*]

[1] Plant Genetics Institute – CNR, UOS FI, Sesto Fiorentino (FI); [2] Department of Agricultural and Forest Economics, Engineering, Sciences and Technologies – University of Florence; [3] Department of Agronomy and Land Management, University of Florence; [4] Institute for Biometeorology, CNR, UOS SS, Li Punti, Sassari; [5] Institute of Ecosystem Study – CNR, UOS FI, Sesto Fiorentino (FI); [6] Settore Promozione dell'innovazione e sistemi della conoscenza, Regional Government of Tuscany, Firenze; [7] Migliarino, San Rossore, Massaciuccoli Regional Park, Pisa; [8] Plant Genetics Institute, CNR, Bari; [9] Environmental Management Agency srl, Buti (PI); Italy

Abstract

In this work we studied the pollen flow of a selected range of crops which could be genetically transformed in the near future. The study is part of LIFE08 NAT/IT/342 DEMETRA project which aims at developing a quick monitoring index to rapidly assess the influence on ecosystems of transgenic crops.

To do this three experimental plots were selected in the Migliarino – San Rossore – Massaciuccoli Regional Park (Tuscany, Italy). The plots were characterized by different cropped areas: maize, sunflower, oilseed rape, Italian stone pine, and poplar. Pollen traps were installed within the plots taking into account the distance from crops and wind direction. A pollen dispersal simulation model was used to assess the potential contamination levels due to maize crops.

Our results show that maize pollen covered up to a distance of 160 m from the cultivated area; pollen granules of oilseed rape, sunflower, pine and poplar were detected up to a distance of 34 m, 19 m, 269 m and 380 m, respectively. The pollen dispersal simulation model provided spatial explicit estimations of the contamination levels in term of maize pollen granules concentration.

---

* Corresponding author: cristina.vettori@cnr.it

Introduction

Pollen dispersal data are necessary to assess the potential impact on ecosystems of genetically modified organisms (GMOs). For instance, information on pollen flow is needed to assess the distances of transgenic pollen that could affect biodiversity and target species. This knowledge is also necessary to identify potential times and environmental conditions that might favour the extension of large pollen flow distances. Additionally, information on pollen flow can be used to assess the pollen dispersal using dispersal simulation models.

Several studies have analyzed maize pollen flow and deposition using pollen traps (e.g., Brunet et al. 2003; Walklate et al. 2004; Devos et al. 2005) while other have investigated pollen flow distance for poplar using molecular markers (e.g., Burczyk et al. 2004; DiFazio et al. 2004).

Several pollen dispersal simulation models have been proposed in the last years (e.g., Lavigne et al. 1996; Colbach et al. 2001a; Colbach et al. 2001b; Klein et al. 2003; Balducci et al. 2007; Mazzoncini et al. 2007). However, these models do not consider relevant features like elevations and the presence of natural and artificial barriers (e.g., hedges, windbreaks) that can play an important role in pollen filtration/dilution.

Our work focuses on pollen dispersal data of the following species (crops and trees) that might be genetically modified in the near future: maize (*Zea mays* L.), oilseed rape (*Brassica napus* L.), sunflower (*Helianthus annuus* L.), Italian stone pine (*Pinus pinea* L.) and poplar (*Populus nigra x Populus deltoides*). We used pollen traps to assess the distances covered by pollen granules of the selected species and a pollen dispersal simulation model to assess the potential contamination levels due to maize crops. At least to our knowledge, this is the first study that attempt to get data on pollen dispersal for sunflower, oilseed rape (wind and insect pollinated plant) and forest trees such as pine and poplar (wind pollinated trees) using pollen samplers. It is worth noting that even in the case of insect pollinated plants, wind pollination – though often negligible – is a component which should also be analyzed. The selected pollen dispersal simulation model (Gorelli et al. 2008) considers the influence on pollen dispersal due to elevations and barriers, and the data on pollen concentration from pollen traps.

Material and methods

Pollen dispersal data were collected in 2011. Three experimental plots were selected in the Migliarino – San Rossore – Massaciuccoli Regional Park (Tuscany, Italy). The plots were characterized by different cropped areas: maize, sunflower, oilseed rape, Italian stone pine, and poplar. Oilseed rape and pine pollen flow was investigated in plot 1; poplar was analyzed in plot 2; maize and sunflower were examined in plot 3 (Fig. 1). Both pine and poplar were naturally occurring.

A meteorological station was installed in the study area. Air temperature and air humidity were measured at a height of 2 m, while wind speed and wind direction were measured at the height of 2.5 m by means of ultrasonic anemometers. All parameters were recorded every minute and averages were calculated and memorized every 15 minutes. The main wind direction in July 2011 was east-northeast.

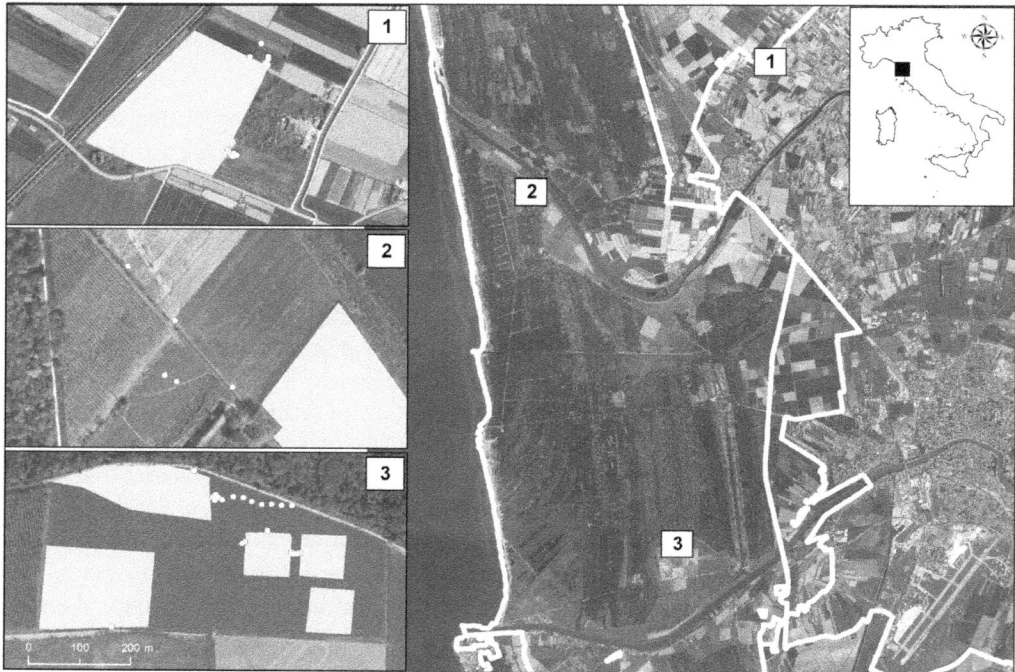

Fig. 1: Study area (right side) and experimental fields in the study plots 1, 2 and 3 (left side: grey polygons represent the cropped areas; white dots represent the pollen traps).

The study on pollen dispersal was carried out to assess the maximum distance at which the pollen can arrive in the study area. To do this, the traps were installed from the border of the cropped areas at increasing distance taking into account the main wind direction during the time of pollen dispersal (Fig. 1). The position of the cropped areas and the position of the traps were recorded using a GPS. The exposure time of pollen samplers was from March 29th to April 26th for oilseed rape and pine, from April 3rd to April 17th for poplar, from July 8th to August 3rd for maize and sunflower. The distances between the pollen traps and the cropped areas are shown in Table 1.

The Sigma-2 pollen samplers were used as they are appropriate for environmental monitoring of GMO (VDI 2007). The Sigma-2 sampler provides a standardized sampling method for direct microscopic pollen analysis, as the pollen adhering to the deposition area is directly analyzed with regard to species and amount by means of light microscopy. An example of the preparation of pollen trap substrates with strips "Silkostrip (Lanzoni srl)" for pollen collection is shown in Figure 2.

Pollen species were differentiated on the basis of their morphological and structural characteristics distinguishing features such as size, form, texture, number of pores, plasma structure, etc. (Fig. 3).

Tab. 1: Pollen traps positioning distance (the distance between the traps and the cropped areas was taken from the edge of the cropped areas).

Crop	Maize												
Pollen Trap	1M	2M	3M	4M	5M	6M	7M	8M	9M	10M	11M	12M	13M
Distance (m)	1	2	3	5	10	20	40	60	80	100	120	140	160
Crop	Oilseed rape												
Pollen Trap	1O	2O	3O	4O	5O	6O	7O						
Distance (m)	1	1.7	3	5.5	13	18	34						
Crop	Sunflower												
Pollen Trap	1S	2S	3S	4S	5S	6S	7S						
Distance (m)	1	2	5	10	13	15	19						
Crop	Italian stone pine												
Pollen Trap	1	2	3	4	5	6	7	8					
Distance (m)	255	256	258	262	263	266	268	269					
Crop	Poplar												
Pollen Trap	1	2	3	4	5								
Distance (m)	70	180	190	240	380								

A   B   C

Fig. 2: A: preparation of substrates with strips "Silkostrip (Lanzoni srl)" for the pollen collection; B: the strips are resting on the bottom of the pollen trap; C: the Sigma-2 pollen samplers ready for use.

Fig. 3: Visual analysis with Optical Microscope.

Photos were taken from each plate. Each of altogether three microscope slides was divided into eight rectangles of 1 cm² (5 cm x 0.2 cm). The pollen granules count was carried out on each of those rectangles. A total of 32 readings per sampling point were carried out. Pollen concentration was computed on the bases of the pollen number visualized on the microscope slides (n. granules/cm²). For each sampling point the pollen concentration was calculated from the readings average (+/− standard deviation).

Data from pollen traps were used to assign a contamination level according to the pollen granules concentration. The contamination levels were used for the implementation of the pollen dispersal simulation model (Gorelli et al. 2008). The model data input was:
a) real-time data: data characterized by a high frequency of updating (e.g. every hour, every day), like precipitation data (mm of rainfall), temperature data (max, min, average), wind data (direction and intensity) and humidity data (relative humidity);
b) off-time data: data characterized by a low frequency of updating (e.g. once a year), like elevations data (e.g., Digital Elevation Model (DEM), Digital Terrain Model (DTM)), data of the barriers' presence (e.g., windbreak, hedges, etc.), and cropped areas (GM and GM-free plots, crop variety, flowering date, etc.).

The pollen dispersal simulation model was applied to assess the impact of GM maize covering 40% of the total cultivation area grown with maize in the study area (19911 ha). We chose the rate of 40% because it is a likely hypothesis of the possible future scenarios as a result of the introduction of GM maize. GM maize fields were chosen randomly.

Results

Pollen dispersion analysis with pollen samplers
In case of maize we found pollen granules up to a distance of 160 m (12 pollen granules/cm² in the farthest pollen trap); the highest decreasing rate of the number of pollen granules occurred between 0 m and 20 m, with 87 to 15 pollen granules/cm². From 20 m up to 160 m the number of pollen granules remained nearly constant (Fig. 4a).

Oilseed rape is an insect as well as a wind pollinated plant. A relevant number of pollen granules of oilseed rape were detected up to a distance of 34 m (49 pollen granules/cm² in the farthest pollen trap). The highest decreasing rate of the number of pollen granules occurred between 0 m and 5 m with 271 to 121 pollen granules/cm². For oilseed rape, the number of pollen granules remains nearly constant up to 18 m and after the value is halved to 49 pollen granules/cm² (Fig. 4b).

Even if sunflower is an insect pollinated we found wind drifted pollen granules up to a distance of 19 m (23 pollen granules/cm² in the farthest pollen trap), with the highest decreasing rate of the number of pollen granules occurring from 0 m to 10 m (from 161 to 18 pollen granules/cm²). In case of sunflower, the number of pollen granules remained nearly constant up to 19 m (Fig. 4c).

The pollen of the Italian stone pine covered up to a distance of 269 m (79 pollen granules/cm^2 in the farthest pollen trap); the highest decreasing rate of pollen granules occurred from 0 m to 262 m. The number of pollen granules remained nearly constant up to 269 m (Fig. 4d). It is worth nothing that the pine is a wind-pollinated tree and the pollen is able to drift over very long distances thanks to its morphology (presence of two wings or air pockets).

For poplar we found a considerable presence of pollen up to a distance of 380 m where the farthest pollen trap was positioned. The number of pollen granules remains nearly constant up to 380 m (about 350 pollen granules/cm^2).

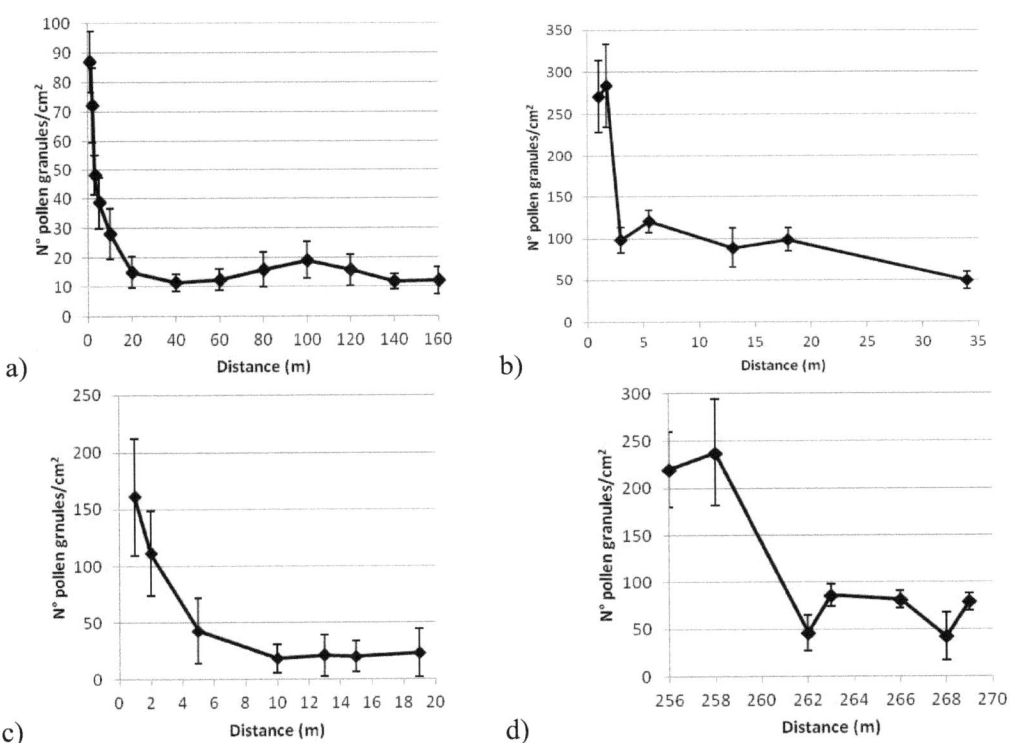

Fig. 4: Pollen flow trend for (a) maize, (b) oilseed rape, (c) sunflower and (d) pine. N° = number of pollen granules/cm2 in relation to the distance from cropped areas. The points are the means ± standard deviation of the means.

Pollen dispersal simulation model
The spatial distribution of maize fields that might be contaminated by the use of GM maize is shown in Fig. 5a. This simulation detected all the plots that could be contaminated by pollen without considering the maximum limit of fall-out of the pollen, therefore, all the target plots affected by the pollen were contaminated. The possible contaminated fields had a total surface of 663 ha, corresponding to 8% of the total area cultivated with non-GM maize.

The data obtained from the pollen traps were then used to assign a contamination level according to the pollen granules concentration: low risk from 0.01 to 31.48 pollen granules/cm^2; medium risk from 31.48 to 94.43 pollen granules/cm^2; high risk from 94.43 to 188.85 pollen granules/cm^2; very high risk from 188.85 to 283.28 pollen granules/cm^2. The risk classes have been identified with respect to the experimental data obtained considering the pollen granules concentration that could potentially affect the contamination level of maize fields. After the implementation of the spatial simulation model with the pollen dispersal data a "risk map" was obtained. The total contaminated fields covered an area of 263 ha, corresponding to 3 % of the non-GM maize cultivated area, and the areas with high (68 ha) and very high (50 ha) risk of contamination were assessed (Fig. 5B). Respect to the first simulation, the total area of contaminated fields is reduced due to the fact that a variable, which takes into consideration the distance from cultivated areas to the pollen emission point, was introduced to model pollen dispersal.

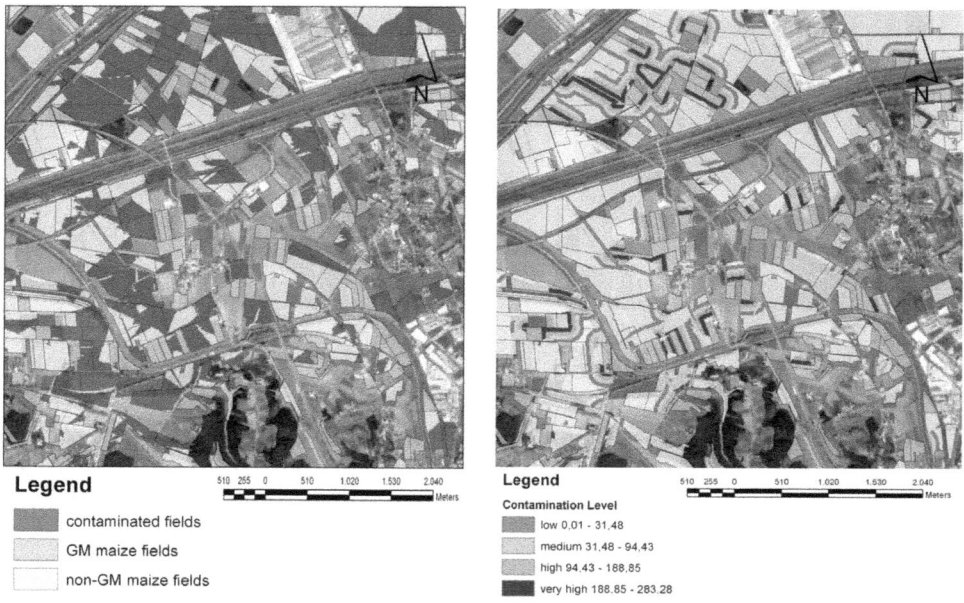

Fig. 5: Simulation results. A) Results obtained without the implementation of the pollen dispersal simulation model using pollen dispersal data; the contaminated area is shown in dark grey color. B) Results obtained with the pollen dispersal data; the contamination levels are shown in different grey colors according to the different risk levels: low, medium, high, very high.

Conclusions

The results achieved with the pollen samplers are essential to collect data concerning the distance covered by transgenic pollen that could impact on biodiversity and target species.

Several studies have been carried out on pollen dispersal of maize considering the measurements of pollen concentrations at various distances and heights from a pollen source. Overall, these studies show that most of the pollen occur within 30–50 m from the source. However, when convective air currents have been considered, the presence of pollen has been observed up to 650 m from a known GM source (Devos et al. 2005).

The distance over which the pollen can be dispersed depends on the local environmental conditions as well as on predominated climatic (wind direction, humidity, temperature, etc.) and it has to be evaluated using a case by case approach. Our results show that the pollen of maize was still detected up to a distance of 160 meters from the cultivated area. Additional studies are in progress to evaluate the viability of the pollen detected at long distance.

Molecular marker have been used for gene flow investigations of oilseed rape, sunflower, pine and poplar (e.g. DiFazio et al. 2004; Damgaard & Kjellsson 2005; De-Lucas et al. 2008; Ureta et al. 2008). For the first time, our study illustrate a pollen dispersion of tree by the wind including the use of pollen traps. For oilseed rape, there are few studies that take into account only the individual plant pollen dispersal by the wind (Lavigne et al. 1998; Klein et al. 2006). This type of knowledge needs to be further investigated in order to limit out-crossing by proposing containment measures also for future GM species that could be posed in commercialization.

Using the data provided by our work, it was possible to implement a pollen dispersal simulation model by identifying different contamination levels (low, medium, high, very high) according to the pollen granules concentration. The spatial simulation model is a tool that can be used to map the effects of the introduction/implementation of different coexistence measures in relation to the level of contamination. It is worth noting that the spatial simulation model can be applied to any crop and environmental context (Gorelli et al. 2008).

Overall, the spatial simulation models allow to take into consideration areas of special interest (e.g., protected areas, Sites of Community Importance, organic farms, areas characterized by typical productions, etc.) as well as identify/evaluate coexistence measures more or less restrictive in connection with the need to preserve such sites from the risk of contamination.

References

Balducci, E., Mazzoncini, M., Gorelli, S. (2007) Coexistence scenarios between GM and GM-free crops. Proceedings of 5th International Conference LCA in Foods, 25-26 April 2007, Gothenburg, Sweden, 103–106.

Brunet, Y., Foueillassar, X., Audran, A., Garrigou, D., Dayau, S., Tardieu, L. (2003) Evidence for long-range transport of viable maize pollen. In: Boelt B. (ed) 1st European Conference on the Co-existence of Genetically Modified Crops with Conventional and Organic Crops. Research Centre Flakkebjerg, 74–76.

Burczyk, J., DiFazio, S. P., Adams, W. T. (2004) Gene Flow in Forest Trees: How far do genes really travel?. Forest Genetics 11: 179-192, 2004.

Colbach, N., Clermont-Dauphin, C., Meynard, J.M. (2001a) GeneSys: a model of the influence of cropping system on gene escape from herbicide tolerant rapeseed crops to rape volunteers. I. Temporal evolution of a population of rapeseed volunteers in a field. Agriculture, Ecosystems and Environment 83: 235–253.

Colbach, N., Clermont-Dauphin, C., Meynard, J.M. (2001b) GeneSys: a model of the influence of cropping system on gene escape from herbicide tolerant rapeseed crops to rape volunteers. II. Genetic exchanges among volunteer and cropped populations in a small region. Agriculture, Ecosystems and Environment 83: 255–270.

Damgaard, C., Kjellsson, G. (2005) Gene flow of oilseed rape (*Brassica napus*) according to isolation distance and buffer zone. Agriculture, Ecosystems and Environment 108: 291–301.

De-Lucas, A.I., Robledo-Arnuncio, J.J., Hidalgo, E., Gonzalez-Martinez, S.C. (2008) Mating system and pollen gene flow in Mediterranean maritime pine. Heredity 100: 390–399.

Devos, Y., Reheul D., De Schrijver, A. (2005) The co-existence between transgenic and non-transgenic maize in the European Union: a focus on pollen flow and cross-fertilization. Environmental Biosafety Research 4: 71–87.

DiFazio, S.P., Slavov, G.T., Burczyk, J., Leonardi, S., Strauss, S.H. (2004) Gene flow from tree plantations and implications for transgenic risk assessment. In: Christian Walter and Mike Carson (eds). Plantation Forest Biotechnology for the 21st Century. Kerala (India), Research Signpost. 405–422.

Gorelli, S., Santucci, A., Balducci, E., Mazzoncini, M., Russu, R. (2008) Spatial simulation model to analyze coexistence scenarios between GM and GM-free crops. Proceedings of the Conference on "Implications of gm-crop cultivation at large spatial scales" April 2–4, University of Bremen, Germany.

Klein, E.K., Lavigne, C., Foueillassar, X., Gouion, P.H., Larédo, C. (2003) Corn pollen dispersal: quasi-mechanistic models and field experiments. Ecological monographs 73: 131–150.

Klein, E.K., Lavigne, C., Picault, H., Renard, M., Gouyon, P.H. (2006) Pollen dispersal of oilseed rape: estimation of the dispersal function and effects of field dimension. Journal of Applied Ecology 43: 141–151.

Lavigne, C., Godelle, B., Reboud, X., Gouyon, P.H. (1996) A method to determine the mean pollen dispersal of individual plants growing within a large pollen source. Theoretical and Applied Genetics 93: 1319–1326.

Lavigne, C., Klein, E.K., Vallée, P., Pierre, J., Godelle, B., Renard, M. (1998) A pollen-dispersal experiment with transgenic oilseed rape. Estimation of the average pollen dispersal of an individual plant within a field. Theoretical and Applied Genetics 96: 886–896.

Mazzoncini, M., Balducci, E., Gorelli, S., Russu, R., Brunori, G. (2007) Coexistence scenarios between GM and GM-free corn in Tuscany region (Italy). Proceedings of Third International Conference on Coexistence between Genetically Modified (GM) and non-GM based Agricultural Supply Chain, 20-21 November 2007 Seville, Spain.

Ureta, M.S., Carrerab, A.D., Cantamutto, M.A., Poverene, M.M. (2008) Gene flow among wild and cultivated sunflower, *Helianthus annuus* in Argentina. Agriculture, Ecosystems and Environment 123: 343–349.

VDI (Verein Deutscher Ingenieure) (2007) Monitoring the effects of genetically modified organisms (GMO) Pollen monitoring Technical pollen sampling using pollen mass filter (PMF) and Sigma-2-sampler. VDI 4330 Part 3. VDI-Handbuch Biotechnologie, Band 1: GVO-Monitoring.

Walklate, P.J., Hunt, J.C.R., Higson, H.L., Sweet, J.B. (2004) A model of pollen - mediated gene flow for oilseed rape. Proceedings of the Royal Society of London. B 271: 441–449.

Can dwarfed Oilseed Rape (*Brassica napus* L.) measure up to tall cultivars?

Jana Seeger, Broder Breckling & Juliane Filser

University of Bremen, UFT Centre of Environmental Research and Sustainable Technology, Bremen, Germany

Abstract

Dwarfing has been proposed as a method of transgenic mitigation to reduce competitive fitness of genetically modified (GM) oilseed rape (OSR) and limit unintended gene flow. But could dwarfing, in contrast, be advantageous for feral plants growing on disturbed ruderal sites? We compared relative fitness of the semi-dwarf hybrid PR45D03 and the tall cultivar Artus in two field experiments, testing the effects of a) low soil quality and b) simulated mowing. PR45D03 produced either a similar amount or less seeds per fruiting plant than Artus under the conditions tested. Our results suggest that dwarfing may reduce the success of feral plants in some situations. Further research needs to test different cultivars and if dwarfing may be advantageous under other conditions.

Introduction

Al-Ahmad et al. (2006) have developed a new transgenic OSR variety carrying transgenes for herbicide resistance and dwarfing. The dwarfed cultivar produces more seeds than tall non-transgenic plants when grown alone, but less when grown in competition. In consequence, transgene escape via OSR volunteers in subsequent crops would presumably be reduced. Yet, concerns have been raised (Reuter et al. 2008) that dwarfing could well be advantageous for feral OSR, which mostly occurs on disturbed sites where competition is scarce and where reduced height could facilitate escape from damage (e.g. through mowing). In addition, dwarfed OSR builds up less vegetative biomass without suffering yield reductions (Wang et al. 2004), which could prove beneficial to plants on low-quality ruderal soils. We therefore hypothesized that a dwarfed cultivar yields more seeds than a tall cultivar on mown plots and on soils with low quality.

Methods

The performance of the non-transgenic semi-dwarf hybrid PR45D03 was compared with the conventional tall cultivar Artus in two experiments. These were conducted on a former dump site, previously used to deposit building rubble, in Bremen, Germany.

Substrate comparison
Plants were grown in containers (Ø 53 cm, open bottom) dug in at ground level and filled to a depth of 34 cm with one of three substrates: 1) humous soil from the dump site, 2) mixed soil (humous soil mixed with sand 1:1) or 3) shallow humous soil (9 cm over building rubble). See Seeger et al. (2010) for soil properties. Each cultivar was sown in two replicate plots per container, resulting in a fully randomized split-plot design with substrate as between-plot and cultivar as within-plot factor. Six blocks with three containers each were set up to obtain twelve replicate plots per substrate/cultivar combination (Fig. 1a).

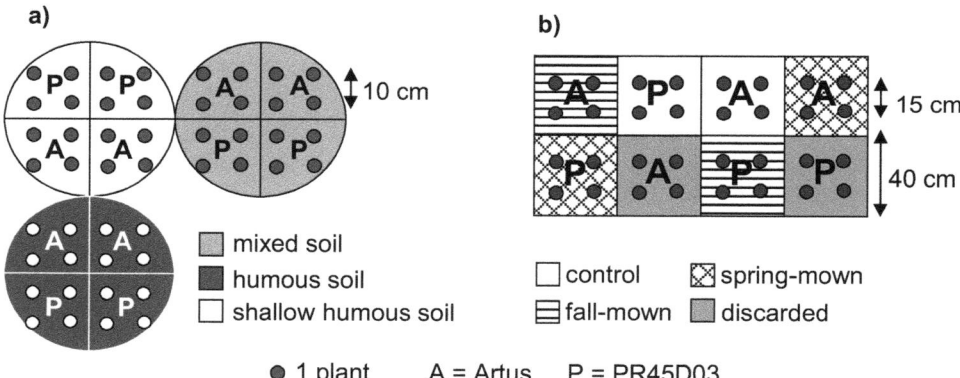

Fig. 1: Set-up of an experimental block. a) Substrate comparison: Each of three containers included two sowing plots per cultivar with 4 plants each. b) Simulated mowing: One block contained four sowing plots per cultivar which were randomly assigned to a mowing treatment.

Simulated mowing
We set up a full two-factorial randomized block design on humous dump site soil with the factors cultivar and mowing regime. Eight blocks were prepared (Fig. 1b), each containing one replicate plot (0.16 m^2) per treatment. Depending on mowing regime, plots were either left unmown, cut to a height of 2.5 cm on 18 December 2008 or cut to 10 cm on 18 April 2009.

Seeds were sown at 1 cm depth on 3 October 2008 and seedlings were thinned to four per replicate plot. Slugs and small mammals were excluded through fences and competing vegetation was regularly removed. The number of fruiting individuals per plot was reduced to averages of 1.7 (substrate comparison) and 2 (simulated mowing) due to high winter mortality. We counted the mean number of pods per fruiting plant and the seeds per pod (based on up to 10 pods per plot) on 14–18 July 2009. The mean number of seeds per fruiting plant was calculated for each replicate plot.

Tab. 1: Seed production per fruiting plant on the three substrates for Artus and the semi-dwarf hybrid PR45D03. Means (bold) and standard errors (SE), n = 7–11.

cultivar	humous		mixed		shallow	
ARTUS	**1860**	405	**702**	132	**1388**	594
PR45D03	**657**	200	**357**	142	**687**	290

Results & Discussion

Substrate comparison
Contrary to our expectations, the tall cultivar Artus produced significantly more seeds per fruiting plant than PR45D03 (Tab. 1, two-way split-plot ANOVA with log-transformed data, $F_{1,55} = 21.30$, $p < 0.001$), regardless of soil quality. While PR45D03 reaches similar yield levels as tall lines in cultivation (Klüßendorf-Feiffer 2009), it appears to be less fit on low-quality soils (see also Sieling & Kage 2008). Possibly, dwarfed varieties profit from their high resistance to lodging (collapse of the plant) only at high N supply (Wang et al. 2004).

Fig. 2: Plant rosettes of Artus (left) and the semi-dwarf hybrid PR45D03 (right) after simulated mowing in fall. The tall cultivar lost more plant biomass.

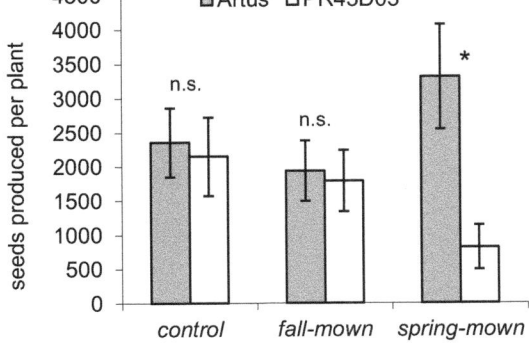

Fig. 3: Seed production per fruiting plant of Artus and PR45D03 on plots with different mowing treatments. Significant (*) differences between cultivars are indicated. Means and standard errors, n = 5-8.

Simulated mowing
Plants of Artus were indeed higher than those of PR45D03 during early growth stages and lost more dry plant biomass at both mowing dates (Fig. 2, data not shown). Nevertheless, Artus yielded as many seeds as PR45D03 on fall-mown plots (Fig. 3, one-way blocked ANOVA n.s.) and even produced significantly more seeds on spring-mown plots (Fig. 3, Welch's ANOVA, $F_{1,9.23} = 9.09$, $p = 0.014$, cultivar effect also significant in two-way blocked ANOVA, $F_{1,42} = 1.44$, $p = 0.025$). Apparently, the tall cultivar showed a higher compensatory ability.

Conclusions

Dwarfing *per se* does not appear to be advantageous for ruderal OSR under the conditions tested, and sometimes reduced seed production. It might therefore mitigate (but not prevent!) gene flow from GM OSR. However, dwarfing may increase yield in cultivation (Al-Ahmad et al. 2006) and therefore likely also on disturbed high-quality ruderal soils. Further research needs to establish if ruderal OSR would profit from dwarfing under other conditions. The effect of sowing date needs to be tested in this context, as late sowing (as in this study) could possibly be tolerated better by tall (i.e. more winter-hardy) cultivars. Mowing experiments need to be repeated with other cultivars and under stressors which could limit the compensatory ability of tall cultivars, e.g. lower mowing height, competing vegetation, low soil quality and herbivore pressure.

Acknowledgements: Seeds were kindly provided free of charge by Pioneer Hi-Bred, Buxtehude, Germany and by RAPOOL-RING GmbH, Isernhagen, Germany. Many thanks to Stephan Hackmann for help with the field work.

References

Al-Ahmad, H., Dwyer, J., Moloney, M., Gressel, J. (2006) Mitigation of establishment of *Brassica napus* transgene in volunteers using a tandem construct containing a selectively unfit gene. Plant Biotechnology Journal 4 (1): 7–21.
Klüßendorf-Feiffer, A. (2009) Druscheignung als zentrale Führungsgröße im Erntemanagement. Dissertation, Berlin, Humboldt-Universität zu Berlin.
Reuter, H., Menzel, G., Pehlke, H., Breckling, B. (2008) Hazard mitigation or mitigation hazard? Environmental Science and Pollution Research 15 (7): 529–535.
Seeger, J., Breckling, B., Filser, J. (2010) Seedling emergence of oilseed rape (*B. napus* L.) and wild relatives on ruderal soils. In: Breckling B, Verhoeven R. (eds) Implications of GM-Crop Cultivation at Large Spatial Scales. Theorie in der Ökologie 16. Frankfurt, Peter Lang. 34–36.
Sieling, K., Kage, H. (2008) The potential of semi-dwarf oilseed rape genotypes to reduce the risk of N leaching. Journal of Agricultural Science 146: 77–84.
Wang, M.L., Zhao, Y., Chen, F., Yin, X.C. (2004) Inheritance and potentials of a mutated dwarfing gene ndf1 in *Brassica napus*. Plant Breeding 123 (5): 449–453.

Chapter II

Landscape effects and agro-ecological interferences

Domestication, feral species and the importance of industrial agriculture to the future of plant diversity

Cynthia Sagers, Meredith Schafer, Brett Murdoch, Jason Londo, Steven Travers & Peter van de Water

University of Arkansas, Fayetteville, Arkansas, USA

> "The evolutionary lines most likely to take advantage of a changing environment are those in which recombination is raised to a maximum. This is accomplished most effectively by mass hybridization between populations having different adaptive norms." (Stebbins, 1959)

Extended Abstract[1]

Stebbins's review of hybridization, evolution and the origins of organic diversity foreshadowed the current era in which cultivated crops commonly grow in the presence of sexually compatible relatives. Co-localization and subsequent hybridization between populations of domesticated and native plant species are increasingly likely as natural landscapes are converted to managed ones. As a consequence, the interface of agricultural and natural systems has tremendous potential for the rapid evolution of new plant forms. Marginal habitats are becoming important repositories of genetic diversity and their study at large spatial scales is increasingly relevant.

In this talk I examine ecological and evolutionary processes that shape patterns of biological diversity at the interface of cultivated and natural systems. I will address a series of topics related to the migration of crop alleles including the movement of transgenes out of agricultural fields, the effects of hybridization on populations of native species, and the preservation of existing diversity in marginal areas that border on managed landscapes.

Large-scale analysis of the risks of crop escape

We are using geospatial tools to more clearly identify risks linked to the dispersal of transgenes into marginal habitats. Geospatial approaches that combine the occurrence of escapees, plant physiological status and environmental parameters will provide a predictive model of the likelihood of persistence of plants escaped from cultivation. In one

---

1 A full paper is in preparation.

project we are adopting stable isotope analyses to assess plant stress across a range of environmental variables. In a further effort, we will use wind vector modelling to design survey routes for the detection of GM plants throughout midwestern U.S. A geospatial perspective allows us to evaluate conditions that favor GM crop escape and to gauge the risks of long-term persistence at a scale comparable to Wilkinson et al. (2003).

References

Kareiva, P., Watts, S., McDonald, R., Boucher, T. (2007) Domesticated nature: shaping landscapes and ecosystems for human welfare. Science 316:1866–1869.

Sarukhán, J., Whyte, A. (2007) Millennium Ecosystem Assessment, Ecosystems and human Well-Being: Current State and Trends. Island Press, Washington, DC.

Stebbins, G.L. (1959) The role of hybridization in evolution. Proceedings of the American Philosophical Society 103:231–251.

Wilkinson, M.J., Elliott, L.J., Allainguillaume, J., Shaw, M.W., Norris, C., Welters, R., Alexander, M., Sweet, J., Mason, D.C. (2003) Hybridization between *Brassica napus* and *B. rapa* on a national scale in the United Kingdom. Science 302:457–459.

Breckling, B. & Verhoeven, R. (2013) GM-Crop Cultivation – Ecological Effects on a Landscape Scale. Theorie in der Ökologie 17. Frankfurt, Peter Lang.

Large scale and small scale approaches for assessing potential exposure of habitats and species neighbouring GM plant cultivation

Frieder Graef[1], Anne Heyer[1], Sigrid Ehlert[1], Ulrich Stachow[1], Claudia Bethwell[1], Sarah Effertz[2], Klaus Henle[2] & Birgit Winkel[3]

[1]ZALF – Leibniz Centre for Agricultural Landscape Research, Müncheberg; [2]UFZ – Helmholtz Centre for Environmental Research, Leipzig; [3]BfN – Federal Agency for Nature Conservation, Bonn; Germany

Abstract

According to EU Directive 2001/18 both environmental risk assessment and monitoring of genetically modified plants have to be carried out for cultivation areas and neighbouring habitats and should be able to detect site-specific and/or regional differences. We present two methodologies complementing one another to determine the frequency and probability of typical and protected habitats and species in the vicinity (50 m, 200 m, 1000 m) of arable land in German agricultural landscapes. We analyse spatial map information of different types and scales in order to assess the potential exposure of habitats and species to GM cropping. We use two analytical approaches differing in methodology, geo-data used and scale applied. On large scales more generalised habitat and species information are analysed, while on small scales the more detailed variability and dynamics of cultivation systems and neighbouring habitats and species are integrated.

Introduction

Possible environment risks of genetically modified plants (GMP) according to European legislation need to be assessed prior to their release and marketing and monitored after release and during cultivation (European Commission 2001). The exposure to GMP primarily encompasses the cultivated fields and neighbouring environment, both fields and other habitats. However, current environmental risk assessments (ERA) and monitoring neglect the variability of exposure intensities which may depend on regional landscape specifics, GM crop source distances, GM crop species, neighbouring habitats and species, and the agricultural practice (Hilbeck et al. 2011).

For large scale ERAs and monitoring the more general information and/or geo-data on landscape, region and climate is relevant (EFSA 2010). Information on specific habitats and species on this scale is usually rather imprecise. For small scale ERAs and monitoring Bethwell et al. (2012) indicate that more detailed information on habitats and species can be used if available in-depths investigations exist. So, the variability and dynamics of agricultural cultivation and crop rotation systems and neighbouring habitats and species can be integrated.

For both large and small scales we develop and present approaches for assessing the potential exposure (co-incidence) of GM cropping and non-target species and habitats, both either protected or common, and we assess options for combining these approaches.

Methodological approaches

An approximation for assessing the potential exposure of habitats and species to GM cropping is information on the probability and frequency of (frequent and/or protected) species and habitats occurring in the vicinity of arable land. The vicinity can be regarded as a function of a) crop species types with inherent spatial effect distances (e.g. pollen range), b) the occurrence (plants) and/or activity radius (animals) of non-target organism species, and c) specific related cause/effect relationships such as the insecticidal effect of the Bt-Toxin on Lepidoptera living in habitats adjacent to GM fields. Hence, to cover a range of different possible GM crops, their potential vicinity to species and possible cause/effect relationships we selected three buffer distances, 50 m, 200 m, and 1000 m.

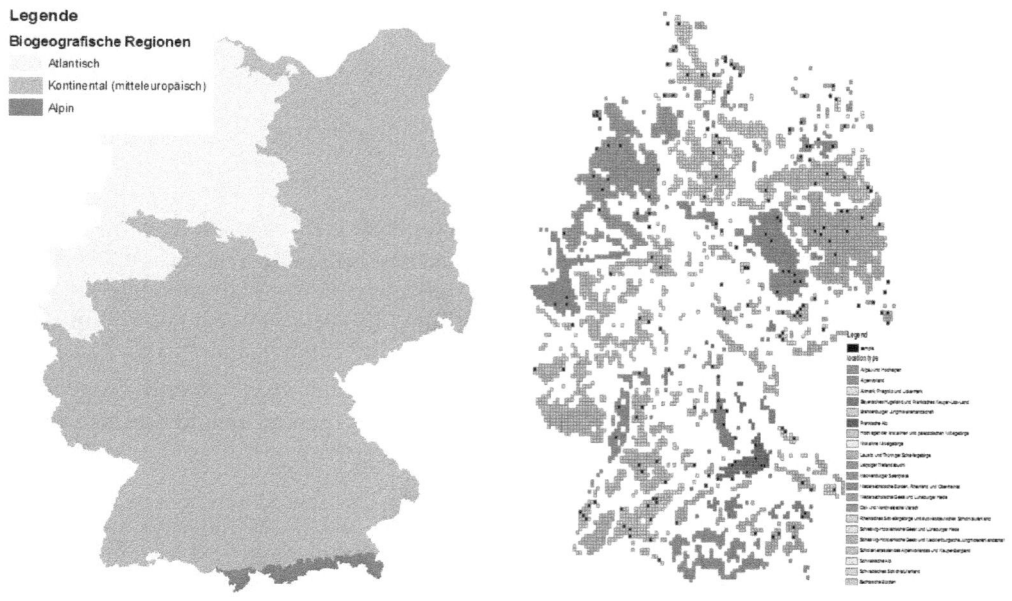

a) EU biogeografic regions   b) Ecoregions (randomly sampled 5x5 km squares)

Large scale and small scale approaches for assessing potential exposure ... 63

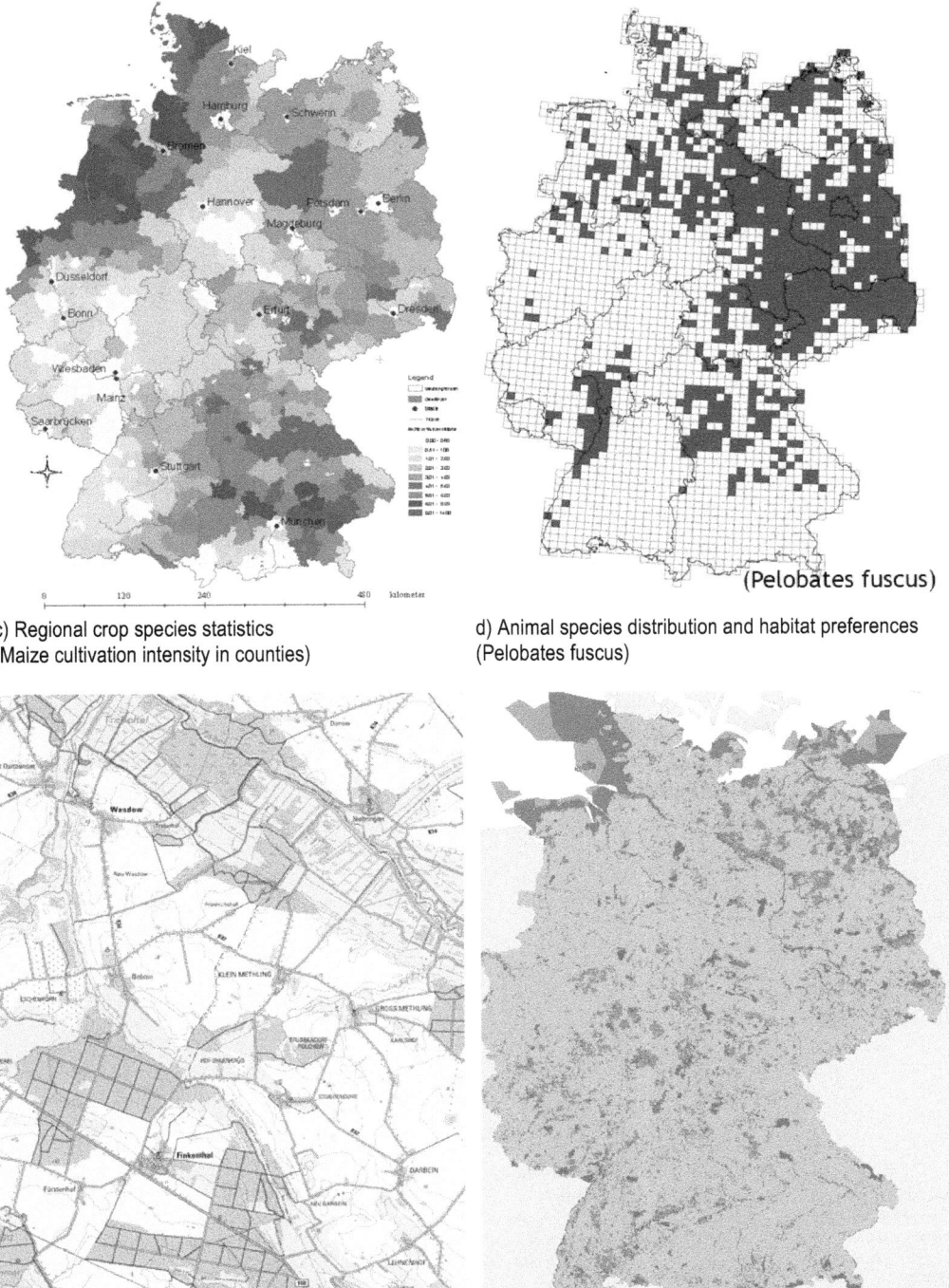

c) Regional crop species statistics
(Maize cultivation intensity in counties)

d) Animal species distribution and habitat preferences
(Pelobates fuscus)

e) ATKIS (Authoritative Topographic Cartographic
Information System) Regional crop species statistics

f) NATURA 2000 and other national protected areas

Fig. 1 a-f: Map information for large scale ERA and environmental monitoring.

For large scale (Germany-wide) ERAs and monitoring and to account for regional ecological and land use differences we analysed ecoregion maps of Germany (Schröder & Schmidt 2001), EU biogeographic regions (EEA 2012), and ATKIS (Authoritative Topographic Cartographic Information System) data (AdV 2012), CIR (Color infrared) biotope and land use maps (BfN 2012), maps of protected biotopes and species (BfN 2012), NATURA 2000 information (BfN 2012), national species occurrence maps from 32 bird, reptile, amphibian and insect species (Ackermann et al. 2012), and regional crop species statistics (Fig. 1). To be able to analyse the extremely extensive ATKIS and biotope data within the 200 m and 1000 m distance, we randomly selected 180 squares of 5 x 5 km among a total of 18 ecoregion classes with arable fields (Fig. 1b).

For small scale ERAs and monitoring we used field maps and related field-specific annual cultivation data for four individual farms selected to be representative for Brandenburg, Germany, and CORINE Land Cover data for characterising their environment (Bethwell et al. 2012). We assumed that all maize grown is GM (Bt-Maize) and calculated the temporal exposure frequency among six years (2002 - 2007) for Bt-Maize field patches and neighbouring patches, either fields or other habitats (Fig 2). Hence, we are able to identify sites that are more often and intensively exposed to Bt-maize and prioritize areas for environmental monitoring or ERAs.

Applying to both large and small scale, within the three buffer distances around arable fields we extract and statistically analyse the various categories of habitats and land uses. Selected wildlife animal species can be systematically assigned to habitats if their habitat preferences are known and statistics can be generated about their likelihood of presence in the vicinity of GM cropping. To prove the feasibility of our approach we select 32 bird, reptile, amphibian and insect species (Ackermann et al. 2012), according to specific criteria and priorities, that we applied sequentially in the following order:

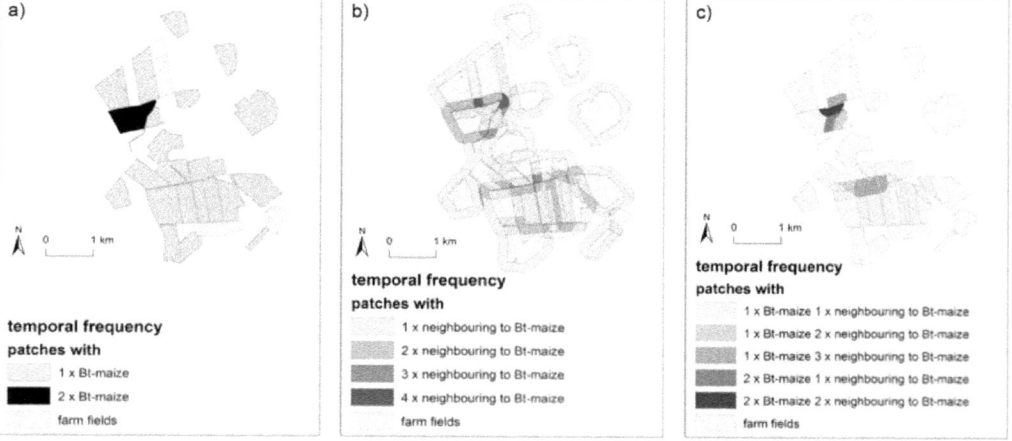

Fig. 2: (a) exposure to Bt-maize fields; (b) exposure to patches neighbouring Bt-maize cultivation (200 m); (c) patches with Bt-maize cultivation and consecutive exposure from neighbouring Bt-maize fields (200 m); period 2002-2007 (Bethwell et al. 2012).

1. frequency and distribution maps of species;
2. information on species' habitat preferences;
3. coverage of different animal classes;
4. differing ecological preferences;
5. national responsibility for the conservation of species.

Both biotope and species information can be further refined and/or attributed to specific GM crop species cultivation regions, ecoregions, and if required to administrative units such as the German Länder.

Results and conclusions

Both the large and the small scale approach provide information on areas with differing exposure levels and can thus be used for prioritising sites for in-depth investigation in the ERA and the environmental monitoring. Large scale analyses are suitable for nationwide region-specific ERAs and the setup of monitoring designs, while small scale analyses/assessments are suitable for detailed field site selections and detection of cause/effect relations on farm level. Since ERAs and environmental monitoring should consider different spatial scales when investigating for adverse effects (Graef et al. 2010) we suggest that both large and small scale assessment approaches should be used and complement each other in a combined integrated framework.

Since the sites are prioritised using a transparent sequential scheme, both the large and the small scale approach yield an improved and objective spatial ERA and monitoring design. This provides detailed exposure information that can be transferred to any agricultural setting.

References

Ackermann, W., Balzer, S., Ellwanger, G., Gnittke, I., Kruess, A., May, R., Riecken, U., Sachteleben, J., Schröder, E. (2012) Hot Spots der biologischen Vielfalt in Deutschland. Auswahl und Abgrenzung als Grundlage für das Bundesförderprogramm zur Umsetzung der Nationalen Strategie zur biologischen Vielfalt. Natur und Landschaft 87 (7): 289–297.

AdV (Arbeitsgemeinschaft der Vermessungsverwaltungen der Länder der Bundesrepublik Deutschland) (2012) http://www.adv-online.de

BfN (Federal Agency for Nature Conservation) (2012) http://www.geodienste.bfn.de/schutzgebiete

Bethwell, C., Müller, H.-J., Eulenstein, F., Graef, F. (2012) Prioritizing GM crop monitoring sites in the dynamics of cultivation systems and their environment. J. Environ. Monitoring 14: 1453.

EFSA (European Food Safety Authority) (2010) Guidance on the environmental risk assessment of genetically modified plants, EFSA Journal 8 (11): 1879.

EEA (European Environmental Agency) (2012) http://www.eea.europa.eu/data-and-maps/figures/biogeographical-regions-europe-2001

European Commission (2001) Directive 2001/18/EC of the European Parliament and of the Council of 12 March 2001 on the deliberate release into the environment of genetically modified organisms and repealing Council Directive 90/220/EEC (Luxembourg, Publications Office) Official Journal of the European Communities, 44 (L106): 1–39.

Graef, F., Schütte, G., Winkel, B., Teichmann, H., Mertens, M. (2010) Scale implications for environmental risk assessment and monitoring of the cultivation of genetically modified herbicide-resistant sugar beet: A review. Living Rev. Landscape Res. 4: 3. http://www.livingreviews.org/lrlr-2010-3

Hilbeck, A., Meier, M., Römbke, J., Jänsch, S., Teichmann, H., Tappeser, B. (2011) Environmental Risk Assessment of Genetically Modified Plants – Concepts and Controversies. Environmental Sciences Europe 23: 13.

Schröder, W., Schmidt, G. (2001) Defining ecoregions as framework for the assessment of ecological monitoring networks in Germany by means of GIS and classification and regression trees (CART). Gate Environmental and Health Sciences 3: 1–9.

Coexistence in Maize: Efficacy of non-GM border rows in reducing pollen-mediated gene flow

Maren Langhof & Gerhard Rühl

Institute of Crop and Soil Science, Julius Kühn-Institut, Braunschweig, Germany

Abstract

A non-GM maize border directly at the edge of the field grown with GM maize has been considered as coexistence measure and already found its way into national coexistence regulations of several EU member states. In large scale field experiments we tested the efficacy of 9 m and 18 m wide recipient maize borders in combination with different isolation distances. In 2008, we combined non-GM maize borders with an isolation distance of 51 m in three sites in Germany. We could not observe any effect up to 18 m wide non-GM maize borders on pollen-mediated gene flow. In a modified field trial conducted in 2010 to 2012 the isolation distance between donor and recipient field was reduced to 6 m and 12 m, respectively, to assess the efficacy of border rows as coexistence measure for small structured agricultural landscapes. In these trials maize gene flow was investigated using a GM-free test system based on the trait kernel color. Results obtained point out that both, 9 m and 18 m wide recipient maize borders do not reliably reduce outcrossing rates in the whole field's harvest regardless of the tested isolation distance (6, 12, 51 m).

Introduction

Isolation distances, buffer crops, separate harvest or shifted sowing dates are considered as possible measures to enable the coexistence between GM and non-GM maize. The efficacy of these measures in reducing outcrossing has been investigated in several studies (Langhof & Rühl 2008). Though many times accepted as coexistence measure, up to now, the efficacy of non-GM maize border rows for reducing pollen-mediated gene flow has not been scientifically verified by appropriate field experiments. Therefore, the aim of this study is to assess the efficacy of 9 m and 18 m wide recipient maize borders in combination with isolation distances of 6 m and 12 m, respectively, in practice-oriented field trials.

Materials & Methods

The field experiments were established at the two sites Braunschweig and Mariensee in 2010 and were repeated at the site Mariensee in 2011 and 2012. Due to the ban on cultivation of GM maize varieties in Germany, the yellow kernelled maize variety Delitop was used as donor and the white kernelled variety DSP 17007 as recipient. To synchronize flowering of donor and recipient maize, cultivars of a similar ripening group were used and sown on the same day. Figure 1 shows the study design. Each recipient field (size 80 x 100 m, white kernelled maize) was separated from the donor field (size 80 x 100 m, yellow kernelled maize) by an isolation distance of 6 and 12 m, respectively. In addition, 9 and 18 m wide borders of white-kernelled maize were established at the edges of two of the donor fields, whereas no border was planted at the third donor field, i.e. the control plot. The areas between donor and recipient maize were planted with a clover-grass mixture.

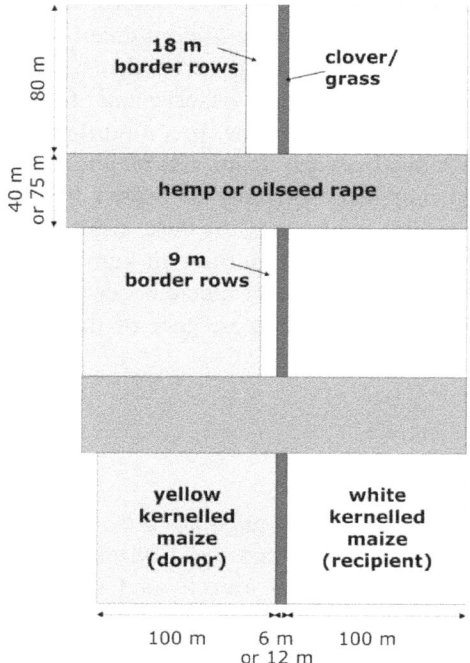

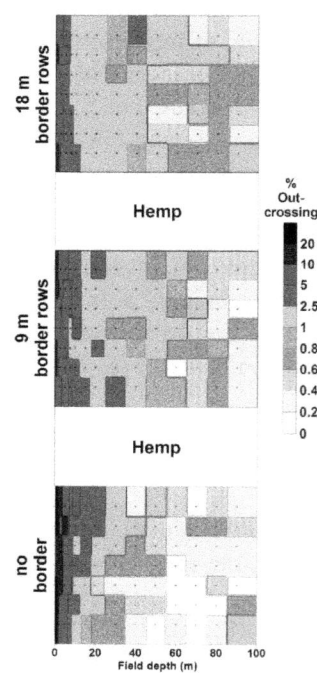

Fig. 1: Sketch of the experimental field design established in 2010 to 2011. Each recipient field (white maize) was separated from the donor field (yellow maize) by an isolation distance of 6 and 12 m, respectively. In addition, 9 m and 18 m wide borders of white kernelled maize were established at the edges of two of the donor fields, whereas no border was planted at the third donor field, i.e. the control plot. The areas between donor and recipient maize were planted with a clover-grass mixture. 75 m wide oilseed rape strips (2010) or 40 m wide hemp strips (2011) were used to isolate the experimental variants from each other.

Fig. 2: Contour plot showing the distribution of pollen-mediated gene flow in the recipient field plots and calculated outcrossing rates in the total harvest exemplarily for the year 2011, field layout with 6m isolation distance between donor and recipient maize.

Tab. 1: Calculated outcrossing rates (%) in the total harvest of the recipient plots in dependence on isolation distance and border row width in 2010 and 2011.

Year	Isolation distance	Border row width		
		0 m	9 m	18 m
2010	6 m	1.8%	0.8%	0.6%
	12 m	0.9%	0.7%	1.2%
2011	6 m	1.8%	1.6%	1.5%
	12 m	2.1%	1.2%	1.2%

Two different crops were used to isolate experimental variants from each other. In 2010, variants were separated from each other by 75 m wide oilseed rape strips. Oilseed rape was planted between variants due to limited space availability at the experimental site. In 2011, 40 m wide hemp strips were used to isolate the experimental plots from each other. In addition to distance separation, hemp plants are expected to prevent pollen exchange between variants because of their dense growth and height of up to 4 m.

Both, donor and recipient maize were grown at a plant density of 90,000 plants/ha and a row spacing of 0.75 m. All maize seeds were protected against soilborne pathogens by standard seed treatment fungicides and bird repellent. Fertilizer and post-emergence herbicides were applied according to regional recommendations and taking soil tests into account. No insecticides were applied. Meteorological data were recorded using on-site weather stations by the German National Meteorological Service (DWD). At each site flowering stages of both maize pollen donor and recipient were recorded at several sampling points during the entire flowering period. Since yellow-kernel color is dominant over white-kernel color numbers of yellow kernels developing on white-kernel maize ears were counted. The percentage of yellow kernels in relation to the mean total kernel number of a white-kernel maize ear was calculated. Outcrossing data were gridded and contoured with the program Surfer version 8.00 with the nearest neighbor method. This gridding method assigns the value of the nearest point to each interpolated point. Using the nearest neighbor method the GM contents in the total harvest of the whole non-GM field were calculated based on measured outcrossing data. Refer to Figure 2 for a contour plot example. For further details on sampling strategy, sample and data analysis refer to Langhof et al. (2008).

Results & Discussion

During maize flowering in 2010 and 2011 synchrony between silking in the recipient and anthesis in the donor field was observed at each site. Temperatures and precipitation during maize flowering represented approximately the long-term means.

The outcrossing rates in the total harvest of the recipient plots are shown in Table 1. Results of the first two experimental years are contrary. In 2010, only in the variant combining an isolation distance of 6 m with a white kernelled maize border of 9 or 18 m depth, the total fields' outcrossing rate was reduced whereas in case of 12 m isolation distance no effect of the maize border was observable. The 18 m border row variant even led to higher outcrossing rates than the 9 m maize border row variant and the control, respectively (Tab. 1). A contrary trend was observed in 2011, only in the variant with 12 m isolation distance the total fields' outcrossing rate was reduced considerably (control: 2.1 %, 9 m and 18 m border row: 1.2 %, respectively). In that year, nearly no effect of the border rows on outcrossing rates in the total harvest of the recipient plots was observed with an isolation distance of 6 m (Fig. 2). Current year results will be available in autumn.

Until now no further systematic studies in efficacy of non-GM maize border rows exist. Our results suggest that not the pollen of the first 18 m of a field is mainly responsible for outcrossing in a neighbouring field but rather the pollen of central parts of a field. The missing border row effect is supported by the hypothesis that pollen of plants at the field edge might be transported over larger distances. Kuparinen et al. (2007) report that large amounts of maize pollen at a field edge travel relatively high and thus disperse over larger distances. Moreover, due to thermal updrafts occasionally a considerable amount of pollen can be transported over long distances (Kuparinen et al. 2007; Hofmann 2008). These assumptions might explain why border rows did not reliably reduce pollen-mediated gene flow in this study.

Conclusions

Planting 9 or 18 m wide conventional maize border rows at the GM field edge is no reliable coexistence measure, at least if combined with an isolation distance of 6, 12 or 51 m. Our results suggest that the separate harvest of the non-GM field edge facing the neighbouring GM field is a more efficient coexistence measure than growing non-GM maize border rows at the edge of the GM field (Langhof et al. 2011).

References

Hofmann, F. (2008) Modellrechnungen zur Ausbreitung von Maispollen unter worst-case-Annahmen mit Vergleich von Freilandmessdaten. http://www.bfn.de/fileadmin/MDB/documents/service/Hofmann_et_al_2009_Maispollen_WorstCase_Modell.pdf.
Kuparinen, A., Schurr, F., Tackenberg, O., O'Hara, R.B. (2007) Air-mediated pollen flow from genetically modified to conventional crops. Ecological Applications 17: 431–440.
Langhof, M., Rühl, G. (2008) Auskreuzungsstudien bei Mais: Überblick, Bewertung, Forschungsbedarf. Berichte über Landwirtschaft 86(1): 29–67.
Langhof, M., Hommel, B., Hüsken, A., Schiemann, J., Wehling, P., Wilhelm, R., Rühl, G. (2008) Coexistence in maize: Do non-maize buffer zones reduce gene flow between maize fields? Crop Science 48: 305–316.
Langhof, M., Hommel, B., Hüsken, A., Mastel, K., Schiemann, J., Wehling, P., Rühl, G. (2011) Coexistence in maize: Effect on pollen-mediated gene flow by conventional maize border rows edging genetically modified maize fields. Crop Science 50: 1496–1508.

Breckling, B. & Verhoeven, R. (2013) GM-Crop Cultivation – Ecological Effects on a Landscape Scale. Theorie in der Ökologie 17. Frankfurt, Peter Lang.

Maize gene flow simulation for intensively used agrarian areas in Lower Saxony (Germany)

Markus Ernsing[1], Broder Breckling[1,2], Hauke Reuter[3] & Gunther Schmidt[2]

[1] Center for Environmental Research and Sustainable Technology (UFT), University of Bremen; [2] Chair of Landscape Ecology, University of Vechta; [3] Leibniz Centre for Tropical Marine Ecology (ZMT), Bremen; Germany

Introduction

The counties of Vechta and Cloppenburg (Lower Saxony, Germany) are located in a highly intensified agricultural environment with one of the highest shares of animal production and maize cultivation in Germany. Compared with Eastern Germany or other intensively cultivated regions the average field size is rather small. If genetically modified (GM) maize would be grown in this agrarian environment, it would be very important to anticipate the efficiency of segregation measures to prevent gene flow from GM to conventional fields. In Germany, there is a segregation distance of 150 m implemented between GM maize and conventional maize. To organic fields, a distance of 300 m has to be kept (GenTPflEV 2008). When the GM impurities in conventional crop go beyond 0.9 %, the conventional harvest must be labelled and marketed as containing GMO (EU 2003). Those regions with a considerable high rate of maize cultivation and relatively small field sizes would be particularly relevant for coexistence feasibility estimates (Ernsing 2012).

Methods

Maize gene flow on the regional level was estimated using the maize dispersal model MAMO (Reuter et al. 2008). To obtain data input for the model we mapped the maize fields in the area north of the city of Vechta and in the surroundings of the city of Friesoythe (county of Cloppenburg). An area of around 20 km² for each location was surveyed. The results were recorded in a map and digitized using the GIS software ArcView 3.3 by ESRI (Liebig & Schaller 2000).

For the estimatation of gene flow, scenarios were defined (Ernsing 2012) with segregation distances of 50 m, 150 m and 300 m, different percentages of GM maize fields (10 %, 40 % and 70 %), and different flowering synchrony (18 days, 6 days and 1 day standard deviation in the onset of flowering).

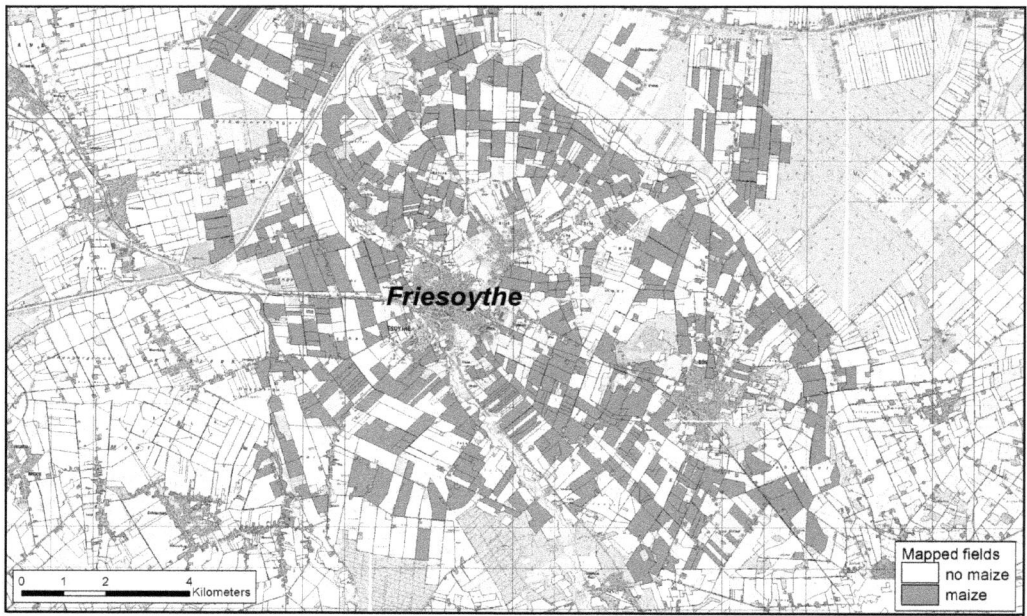

Fig. 1: Maize fields near the city of Friesoythe (county of Cloppenburg) in the year 2011. The grey areas represent maize fields. The other outlined areas were fields with other crops.

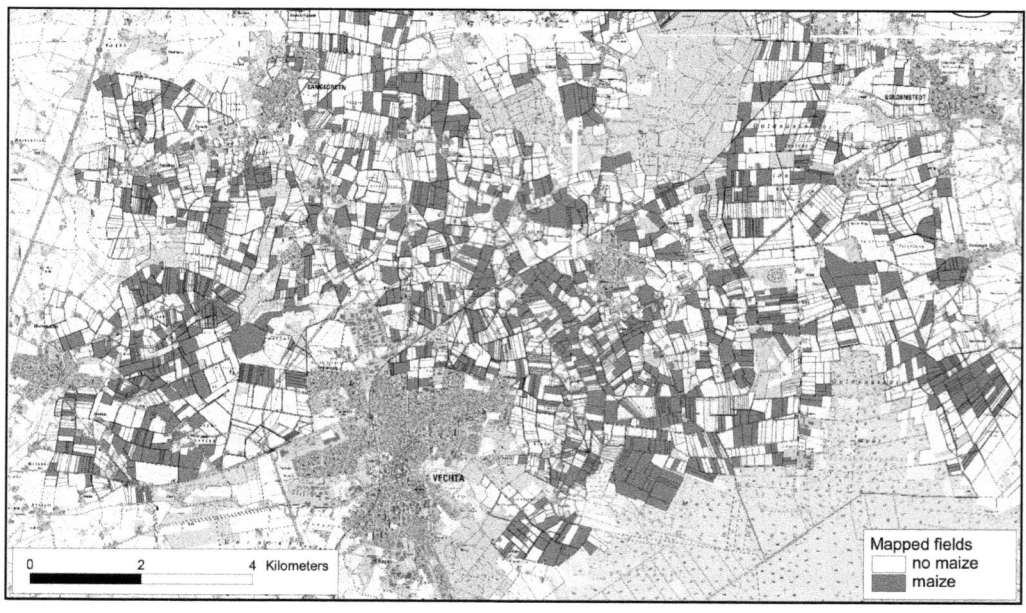

Fig. 2: Maize fields around the city of Vechta in the year 2011. The grey areas represent maize fields, the other outlined areas showed fields with other crops.

Results

Mapping

By mapping, it could be shown that there was a very high density of maize fields in both regions (Fig. 1 and 2). In Vechta, about 30 % of the agricultural area was cultivated with maize, and in Friesoythe about 52 %. The fields surrounding Friesoythe were larger (10.8 ha on average) than the fields in Vechta (2.1 ha on average).

Simulation

The simulation results support the expectation that there is a non-negligible probability of gene flow above 0.9 % for various scenario assumptions. In particular, the contribution of large-distance gene flow was identified to be important. We found that not only the percentage of maize fields with regard to agricultural area and entire area was an important factor for the cross-pollination. Also field size relations played an important role. The highest percentage of gene flow was obtained in the simulations for small conventional fields close to larger neighbouring GM maize fields. Figures 3-7 show the results with a segregation distance of 150 m and with the three different shares of assumed GM maize cultivation.

Friesoythe (county of Cloppenburg)

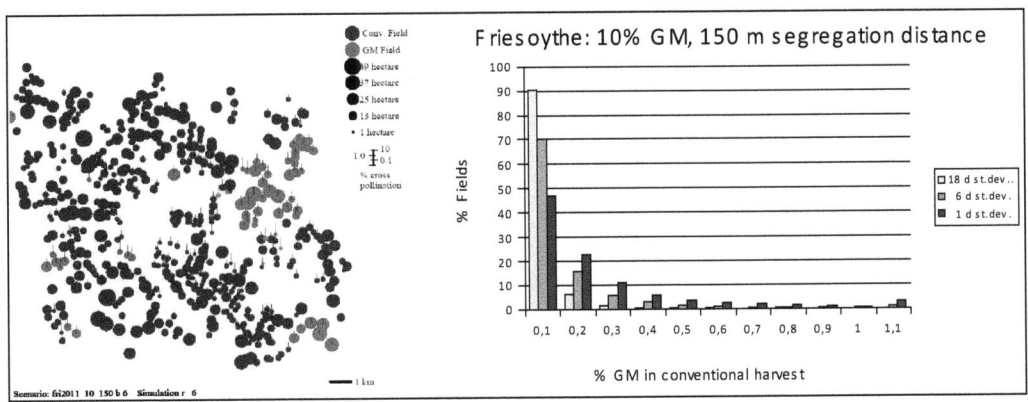

Fig. 3:
Left: Exemplary results of one simulation run for Friesoythe (150 m segregation distance, 10 % GM maize fields and 6 days standard deviation in the onset of flowering). Circles represent fields, vertical bars show the calculated gene flow on a logarithmic scale.
Right: Average regional gene flow for the scenario with 10 % GM maize and a segregation distance of 150 m. The columns indicate the percentage of fields receiving up to 0.1, 0.2 ... 1.0 % of transgenic pollination. The columns 1.1 aggregate all occurrences of higher gene flows rates.

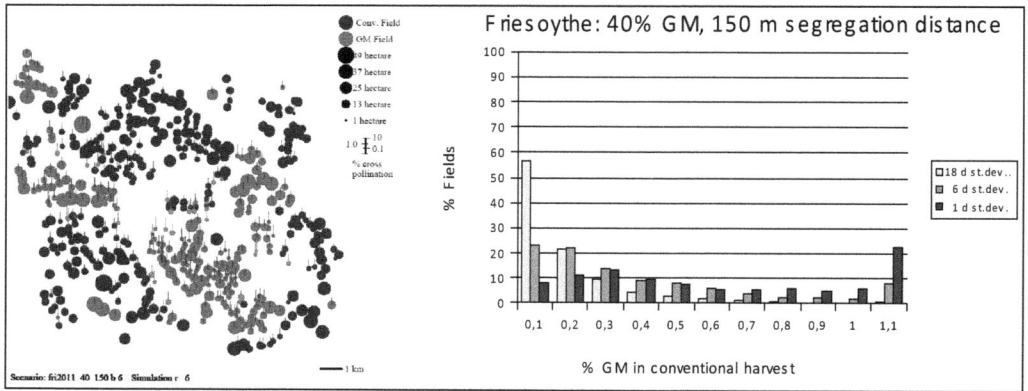

Fig. 4:
Left: Exemplary results of one simulation run for Friesoythe (150 m segregation distance, 40 % GM-maize fields and 6 days standard deviation in the onset of flowering). Circles represent fields, vertical bars show the calculated gene flow on a logarithmic scale.
Right: Average regional gene flow for the scenario with 40 % GM maize and a segregation distance of 150 m. The columns indicate the percentage of fields receiving up to 0.1, 0.2 ... 1.0 % of transgenic pollination. The columns 1.1 aggregate all occurrences of higher gene flows rates.

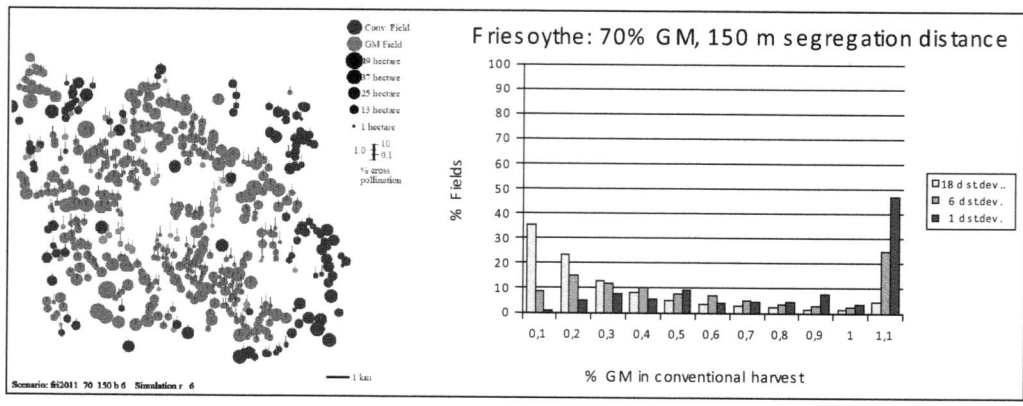

Fig. 5:
Left: Exemplary results of one simulation run for Friesoythe (150 m segregation distance, 70 % GM-maize fields and 6 days standard deviation in the onset of flowering). Circles represent fields, vertical bars show the calculated gene flow on a logarithmic scale.
Right: Average regional gene flow for the scenario with 70 % GM maize and a segregation distance of 150 m. The columns indicate the percentage of fields receiving up to 0.1, 0.2 ... 1.0 % of transgenic pollination. The columns 1.1 aggregate all occurrences of higher gene flows rates.

Vechta

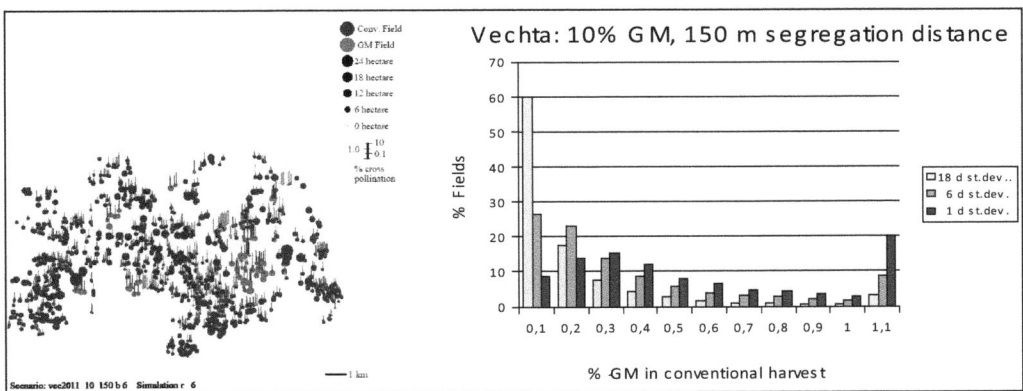

Fig. 6:
Left: Exemplary results of one simulation run for Vechta (150 m segregation distance, 10 % GM-maize fields and 6 days standard deviation in the onset of flowering). Circles represent fields, the vertical bar shows the calculated gene flow on a logarithmic scale
Right: Average regional gene flow for the scenario with 10 % GM- maize and a segregation distance of 150m. The columns indicate the percentage of fields receiving up to 0.1, 0.2 ... 1.0 % of transgenic pollination. The columns 1.1 aggregate all occurrences of higher gene flows rates.

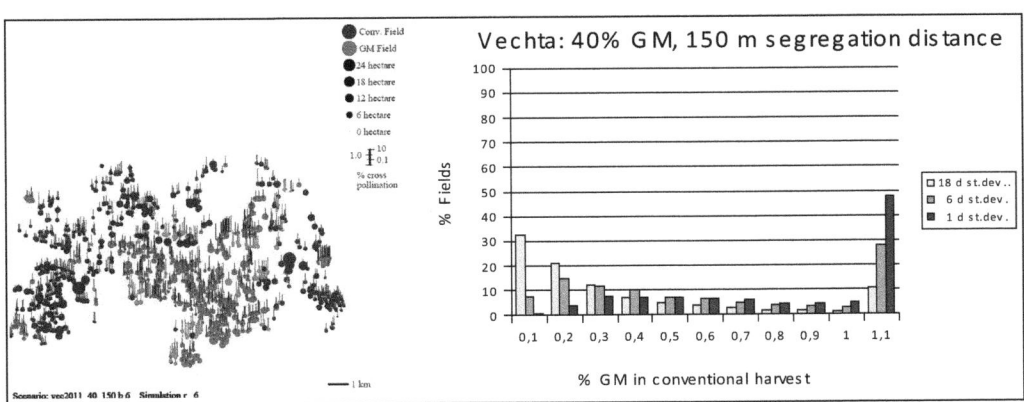

Fig. 7:
Left: Exemplary results of one simulation run for Vechta (150 m segregation distance, 40 % GM-maize fields and 6 days standard deviation in the onset of flowering). Circles represent fields, the vertical bar shows the calculated gene flow on a logarithmic scale.
Right: Average regional gene flow for the scenario with 40 % GM maize and a segregation distance of 150 m. The columns indicate the percentage of fields receiving up to 0.1, 0.2 ... 1.0 % of transgenic pollination. The columns 1.1 aggregate all occurrences of higher gene flows rates.

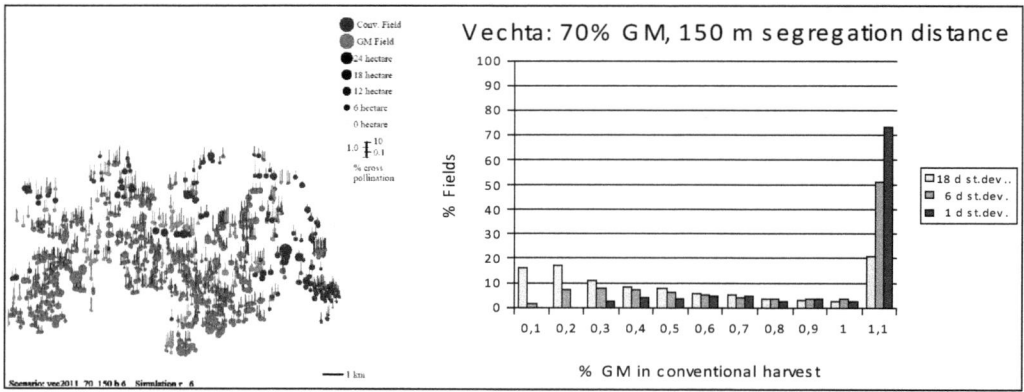

Fig. 8:
Left: Exemplary results of one simulation run for Vechta (150 m segregation distance, 70% GM-maize fields and 6 days standard deviation in the onset of flowering). Circles represent fields, the vertical bar shows the calculated gene flow on a logarithmic scale.
Right: Average regional gene flow for the scenario with 70% GM maize and a segregation distance of 150 m. The columns indicate the percentage of fields receiving up to 0.1, 0.2 ... 1.0% of transgenic pollination. The columns 1.1 aggregate all occurrences of higher gene flows rates.

Conclusions

In areas with high percentages of maize cultivation, gene flow of above 0.9% could occur under various scenario assumptions. A comparison of Vechta and Cloppenburg (Friesoythe) showed that not only the overall cultivation density but also field size and topology are crucial for gene flow estimation. For the simulation of gene flow on the regional scale, estimates of large-distance cross pollination are crucial. Since empirical data are scarce, extrapolations were required. More empirical data on cross-pollination between fields with 0.5 km up to 5 km distance would improve the certainty of estimation. Though the field-to-field gene flow is small across larger distances, the total amount can not be neglected as significantly more neighbours have to be considered.

References

Ernsing M. (2012) Kartierung und Modelluntersuchung von genetisch verändertem Mais in Niedersachsen. Diploma Thesis University of Bremen

EU (2003) Regulation (EC) No 1829/2003 of the European Parliament and of the Council on genetically modified food and feed, Report from the commission to the Council and the European Parliament.

GenTPflEV (2008) Regulations of good agricultural practice at the producing of genetically modified plants in Germany (Gentechnik-Pflanzenerzeugungsverordnung – GenTPflEV). BGBl. I 2008, 658.

Liebig W., Schaller J. (2000) ArcView GIS. GIS-Arbeitsbuch (2nd ed.). Wichmann, Heidelberg, 445 pp.

Reuter H., Böckmann S., Breckling B. (2008) Analysing cross-pollination studies in maize. In: Breckling B., Reuter H., Verhoeven R. (eds.) Implications of GM-Crop Cultivation at Large Spatial Scales. Theorie in der Ökologie 14, Peter Lang, Frankfurt, 47-53.

Breckling, B. & Verhoeven, R. (2013) GM-Crop Cultivation – Ecological Effects on a Landscape Scale. Theorie in der Ökologie 17. Frankfurt, Peter Lang.

Modelling potential maize hybridisation in northern Germany and implementation of a WebGIS application for GMO monitoring issues: Two aspects of the GeneRisk project

Gunther Schmidt, Broder Breckling & Winfried Schröder

Chair of Landscape Ecology, University of Vechta, Germany

Background

The contribution at hand is based on the synthesis report of the *GeneRisk* project (Breckling et al. 2012a) and highlights the modelling of maize hybridization and the WebGIS GVO Monitoring [1]. Additionally, *GeneRisk* key articles were published in the series "Implications for GMO-cultivation and monitoring" (Schmidt & Schröder 2011) launched in the journal *Environmental Sciences Europe* [2].

Genetically modified organisms (GMO) are established in a number of countries, especially in North (USA, Canada: 75.6 Mio. ha in 2010) and South America (Brazil, Argentina: 48.3 Mio. ha in 2010) [3]. Nevertheless, the actual benefits of GMO cultivation are still being discussed controversially. So far, not least because of the restrictive approval policy and due to serious concerns of the consumers, in Europe GMO cultivation plays only a marginal part: In 2010, the entire growing area for GMO was about 91,500 ha, mainly located in the North of Spain. Some countries (e.g., Austria, Hungary, Switzerland and France) even banned GMO cultivation completely because, amongst others, there are concerns about contamination of conventional harvest resulting in expenses for separation of both production chains, evoking liability issues and affecting the consumer's freedom of choice at not least.

Against this background, the *GeneRisk* project, funded from 2006–2010 by the German Federal Ministry of Education and Research (BMBF), was initiated to assess systemic risks of GMO cultivation and to analyse management strategies to cope with potential unintended and undesirable environmental and socio-economic impacts [4]. The *GeneRisk* project investigated systemic risks which could emerge due to large-scale cultivation of GMO encompassing possible ecological and agronomic impacts as well as legal problems and cost-benefit analyses. Accordingly, the *GeneRisk* project consortium evaluated systemic implications of the introduction of GM crops in agriculture by bringing together disciplinary knowledge of ecologists, economists, agronomists, and legal scientists working at the Universities of Vechta, Bremen, Kiel, and Göttingen together with experts from the Leibniz Centre for Agricultural Landscape Research (ZALF) and the Federation of German Scientists (VDW) (Schmidt et al. 2009).

One of the main issues was to assess potential hybridisation rates between GM and conventional maize cropping. For several districts and federal states in Germany, different cultivation scenarios were applied simulating both different shares of GM maize and conventional maize fields and different isolation distances separating them. Additionally, the web-based geographical information system *WebGIS GMO Monitorng* was developed to analyse the feasibility of coexistence regulations with regard to conventional/organic agriculture and the exposition of areas reserved for nature conservation.

Methods

The basic idea of the interdisciplinary approach in the *GeneRisk* project was to structure the assessment of possible GMO effects according to the organisation of hierarchies in the biological sciences, starting with GMO-related modification of molecular interactions and proceed with cells, organisms and interactions on the population level up to ecosystem, landscape and biome-specific interrelations.

Modelling maize hybridisation

Since there are no wild relatives of maize in Central Europe, crossbreeding and establishment of feral hybrid populations is of no concern. However, pollen dispersal by wind and hybridisation of conventional maize has to be considered. Hence, one task in the *GeneRisk* project was to simulate the potential hybridisation rates between conventional and GM maize fields to assess whether the segregation distances prescribed by the German Law on Genetic Engineering (150 m from GM maize fields to conventional maize fields and of 300 m to organic maize fields) would be actually efficient to assure gene flow below the labelling threshold of 0.9 %.

Modelling was realised by the development of the software *MaMo* (Reuter et al. 2008) which based on the programming language SIMULA (Dahl 1968). The gene flow model implemented in *MaMo* was developed as a generally applicable prototype with a specification to assess the cross-pollination (out-crossing) in maize on a regional scale. Different cultivation scenarios were applied simulating both different shares (10 %, 40 %, 70 %) of GM maize and conventional maize fields and different isolation distances separating GMO and conventional maize cultivation.

According to EU regulations, impurities of approved GM varieties are tolerated in conventional crops and in ingredients of food products up to 0.9 %, provided that the impurity occurred unintentional and is technically unavoidable. Above this level, conventional harvest has to be labelled as GMO maize. Compared to local models (Lipsius et al. 2006; Kuparinen et al. 2007; Angevin et al. 2008) explicitly determining the influence of various environmental drivers (wind conditions humidity, topographical patterns etc.), *MaMo* describes the gene flow on a regional scale which lowers the requirements on data availability and accuracy focusing on large scale effects of gene flow (Breckling et al. 2011). *MaMo* uses a dispersal kernel derived by regression analyses of different field studies published in literature and describes the exposition of receptor fields in relation to the distance to adjacent source fields and the resulting hybridisation rate. For each scenario 10 to 15 simulation runs were calculated (Monte Carlo simulation) (Reuter et al. 2008). The variation of the flowering period of all maize fields was considered as well, the standard deviation ranged from 1 to 18 days. The higher this value the lower was the overlapping time for crossbreeding. In combination, 27 different scenarios were calculated for each region.

WebGIS GMO Monitoring
In order to support GMO monitoring and to facilitate coexistence between GMO and conventional/organic crop cultivation and to accomplish nature conservation goals the web-based geoinformation system *WebGIS GMO Monitoring* was established in the course of the *GeneRisk* project [5]. The WebGIS application was developed by using Open Source software exclusively which reduced the costs for the implementation of the system (Kleppin et al. 2011). Only the hardware environment had to be purchased. Open Source software is labelled under the terms of the Open Source Initiative (OSI) (Williams 2002) that defines access, distribution and modification of the software. Examples for licence models of the OSI are the GNU Public License or the Lesser General Public License and BSD License.

When implementing the *WebGIS GMO Monitoring*, standards and specifications for geospatial data and metadata defined by the Open Geospatial Consortium (OGC) and the international standardisation organisation (ISO), like ISO 19115, were considered to ensure interoperability with other map services as indicated by the EU directive 2007/2/EC INSPIRE (Infrastructure for Spatial Information in Europe) [6] and implemented by Geoportal.DE [7] in Germany. The software architecture is based on a combination of the Apache HTTP-Server and the UMN Mapserver working on the operating system Debian Sarge. The UMN Mapserver is capable of reading several data sources like raster and vector data stored in files, databases or integrated into a GIS. The UMN Mapserver meets the standards of the OGC. With the help of the Open Source database management system PostgreSQL in conjunction with the spatial extension PostGIS and the use of the Geometry Engine Open Source (GEOS), a spatial database backend was built up and combined with metadata and other related data. Within this backend, simple GIS functionalities such as buffering and intersecting are provided. Enhanced GIS operations like buffering and clipping were implemented to enable spatial analyses regarding coexistence issues or to allocate environmental observation sites in the vicinity of GMO fields. Visualisation of geodata and interactions between the client and the data are realised by the Mapbender software [8].

Results

Modeling maize hybridisation
The modelling of the maize hybridization rates between GM and conventional maize fields by the *MaMo* model was conducted for different federal states and districts in

Germany reflecting different cultivation patterns and field geometries. For the contribution at hand, we compared the potential gene flow for the federal states of Brandenburg and Schleswig-Holstein in northern Germany. Agriculture in Brandenburg is characterised by a relatively low maize cultivation density, but quite large maize fields, whereas in Schleswig-Holstein complex cultivation patterns with a considerably high number of rather small fields are typical (Breckling et al. 2012b).

Assuming a GMO share of 40 % and a segregation distance of 150 m, at the 6,600 maize fields in Brandenburg unintended gene flow was comparatively low (Fig. 1). For this scenario, the percentage of conventional maize fields potentially being cross-pollinated by GM maize above the labelling threshold value of 0.9 % amounted to 1,25 % in maximum, even if there was a standard deviation in flowering synchrony between the GMO source fields and the receiving conventional maize fields of only one day (i.e. full flowering synchrony).

Assuming a GMO share of 40 % and a segregation distance of 150 m, the 22,000 maize fields in Schleswig-Holstein showed a considerably higher cross-pollination rate (Fig. 2). For this cultivation scenario, the percentage of conventional maize fields potentially being cross-pollinated by GM maize amounted from about 2,8 % to 17,2 % according to the respective flowering synchrony.

WebGIS GMO Monitoring
The *WebGIS GMO Monitoring* contains countrywide geodata describing land use patterns (Corine Landcover) [9], cultivation density of maize on district and municipality level, geometries of nature reserves and Flora-Fauna-Habitats (FFH), and information on environmental monitoring networks. Additionally, for the years 2005 to 2008 all locations GM maize fields (insect resistant Bt-maize) were integrated as well. This information was retrieved from the German GMO location register which is maintained by the Federal Office of Consumer Protection and Food Safety (BVL) [10]. These data were also used to model potential cross-breeding rates between GM and conventional maize fields (section 3.1).

The graphical user interface of *WebGIS GMO Monitoring* consists of 6 elements: (1) map window, (2) reference map, (3) legend, (4) list of available layer, (5) interactive tools, and (6) additional GIS functions (Fig. 3). The toolbox (5) allows simple GIS operations like panning, zooming etc. Enhanced GIS operations (6) were added to perform individual spatial analyses (Kleppin et al. 2011). It is possible to create a buffer around a selected GMO field or a nature reserve to check whether isolation distances are met or to identify all observation sites in the vicinity of a chosen field. For each monitoring site, metadata are defined which can be retrieved from the database by according templates. These metadata include information on measured substances, measuring methods, measuring intervals, and contact details of responsible institutions and persons.

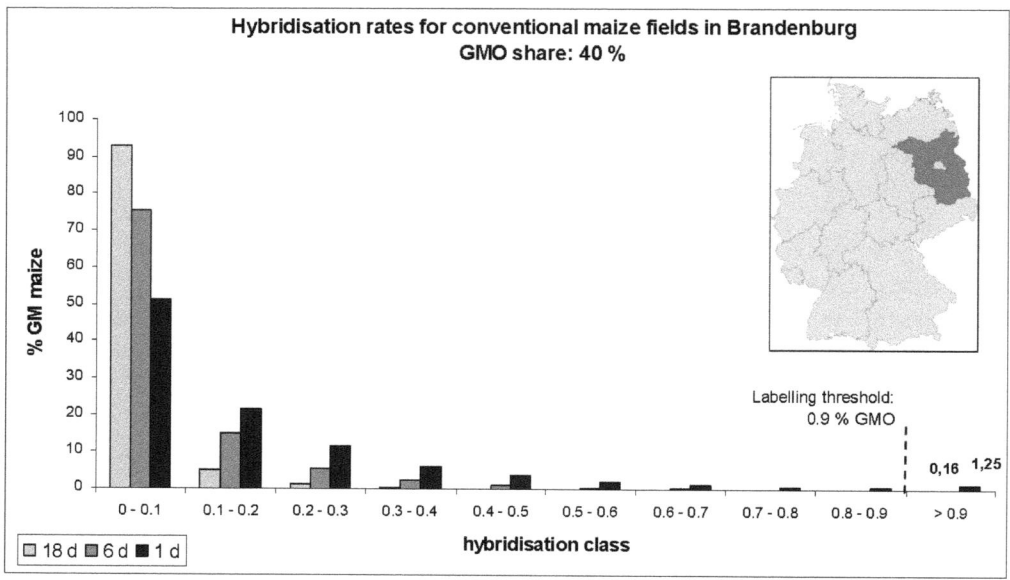

Fig. 1: Modelled mean potential gene flow at conventional maize fields in the federal state of Brandenburg assuming a GMO share of 40%, a minimum distance between conventional and GM maize fields of 150 m and different maize flowering synchrony implemented as a standard deviation of 1 day, 6 days and 18 days in the start of flowering. There were only few conventional maize fields with hybridisation rates of above 0.9% (labelling threshold value), no matter which flowering synchrony was set.

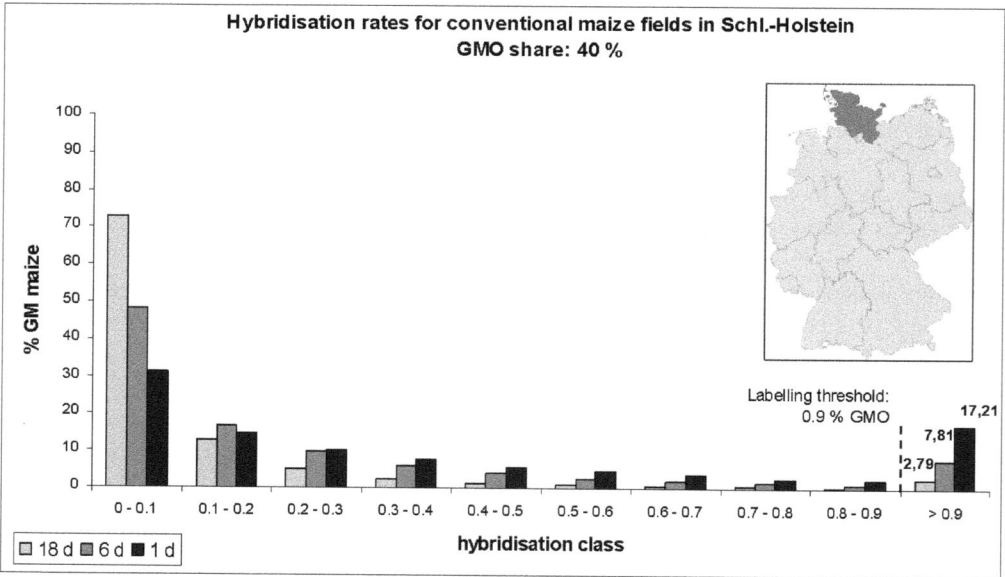

Fig. 2: Modelled mean potential gene flow at conventional maize fields in the federal state of Schleswig-Holstein assuming a GMO share of 40%, a minimum distance between conventional and GM maize fields of 150 m and different maize flowering synchrony implemented as a standard deviation of 1 day, 6 days and 18 days in the start of flowering. There were considerably high percentages of conventional fields with hybridisation rates above 0.9% (labelling threshold), depending on the flowering synchrony between both cropping varieties.

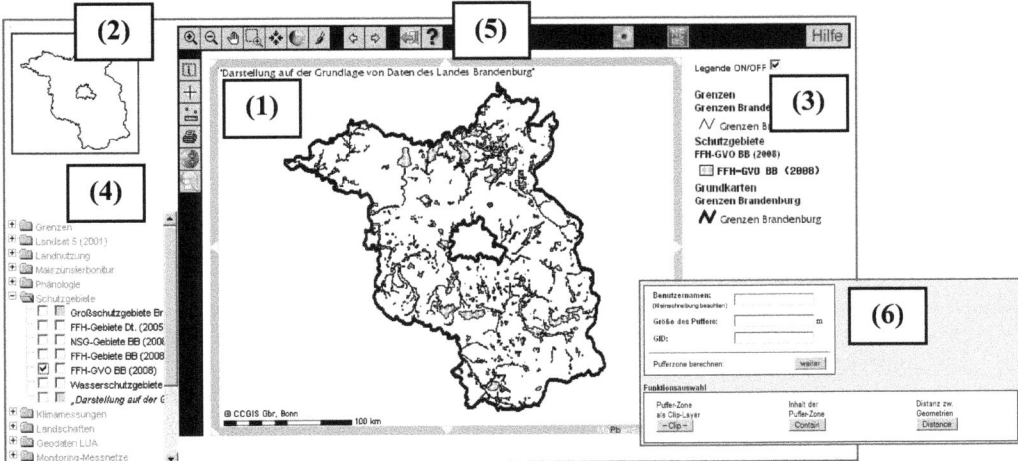

Fig. 3: The user interface of the WebGIS GMO monitoring displaying all FFH areas within the federal state of Brandenburg in its centre. The numbers indicate certain tools of the WebGIS and are explained above.

Conclusions

The modelling results derived by *MaMo* corroborated that under distinct scenario conditions (e.g. field size, topology) higher gene flow between GM and conventional maize fields may occur in particular regions. For example, the hybridisation rates modelled for the 22,000 maize fields in the federal state of Schleswig-Holstein were high when assuming a GMO share of 40 % and a segregation distance of at least 150 m. In contrast, for the given cultivation scenario in the federal state of Brandenburg hybridisation of conventional maize should be negligible since there are less (6,600) but larger maize fields (Breckling et al. 2012b). These findings have relevance, amongst others, when assessing the feasibility of efficient coexistence measures.

The *WebGIS GMO monitoring* is an appropriate tool to link and provide (geo-) information from different sources to facilitate and promote administrative work and public access to GMO related data and information. The WebGIS application gives also support in detecting coexistence problems between GMO cultivation and conventional cropping or nature protection goals already in the planning stage.

Overall, the GeneRisk project illustrated that GMO impact assessment is not sufficiently covered by a purely scientific perspective but requires a comprehensive approach involving ecological or agronomic as well as social science aspects. While the molecular and farm-scale effects of GMO are comparatively well investigated, large-scale effects of GMO are considerably less researched. In this regard, the project contributed the first regional-scale estimation of maize gene flows for Northern German federal states. From an economical point of view, the *GeneRisk* project outlined that – depending on consumer preferences – advantages for seed suppliers and GMO growers can be easily outweight by overall welfare losses due to additional costs for, e.g., environmental monitoring, food inspection, separated food production chains etc. (Barkmann et al. 2010). Specific regulatory and legal questions, e.g. coexistence and liability issues (Stoppe-Ramadan & Winter 2010) or the applicability of modelling approaches (Struss & Winter 2012) have to be considered as well when introducing GMO into the agricultural environment.

References

Angevin, F., Kleinb, E., Choimet, C., Gauffreteau, A., Lavigne, A., Messean, A., Meynard, J. (2008) Modelling impacts of cropping systems and climate on maize cross-pollination in agricultural landscapes: The MAPOD model. European Journal Agronomy 2008. 28:471–484.

Barkmann, J., Thiel, M., Teuvsen, L., Eschenbach, C., Windhorst, W., Marggraf, R. (2010) GM maize and oil seed rape in Germany: Economic welfare losses from large scale adoption scenarios. In: Breckling B, Verhoeven R (eds.): Large-area effects of GM-crop cultivation. Theorie in der Ökologie 16. Peter Lang, Frankfurt, 114–116.

Breckling, B., Schmidt, G., Schröder, W. (Hrsg.) (2012a) GeneRisk – Systemische Risiken der Gentechnik: Analyse von Umweltwirkungen gentechnisch veränderter Organismen in der Landwirtschaft. Springer, Heidelberg.

Breckling, B., Reuter, H., Bethwell, C., Glemnitz, M., Höltl, K., Wurbs, A., Eschenbach, C., Windhorst, W. (2012b) Anwendung des Modells MaMo zur Abschätzung des regionalen Genflusses bei Mais. In: Breckling B, Schmidt G, Schröder W (Hrsg.): GeneRisk – Systemische Risiken der Gentechnik: Analyse von Umweltwirkungen gentechnisch veränderter Organismen in der Landwirtschaft. Springer, Heidelberg, 61–92.

Breckling, B., Reuter, H., Middelhoff, U., Glemnitz, M., Wurbs, A., Schmidt, G., Schröder, W., Windhorst, W. (2011) Risk Indication of Genetically Modified Organisms (GMO): Modelling Environmental Exposure and Dispersal across Different Scales Oilseed Rape in Northern Germany as an Integrated Case Study. Ecological Indicators 11:974–988.

Dahl, O.J. (1968) Simula 67 common base language. Norwegian Computing Centre Publication. Oslo.

Kleppin, L., Schmidt, G., Schröder, W. (2011) Cultivation of GMO in Germany: support of monitoring and coexistence issues by WebGIS technology. Environmental Sciences Europe 23(4), doi: 10.1186/2190-4715-23-2

Kuparinen, A., Schurr, F., Tackenberg, O., O'Hara, R.B. (2007) Air-mediated pollen flow from genetically modified to conventional crops. Ecological Applications 17:431–440.

Lipsius, K., Wilhelm, R., Richter, O., Schmalstieg, K.J., Schiemann, J. (2006) Meteorological input data requirements to predict cross-pollination of GMO maize with Lagrangian approaches. Environmental Biosafety Research 5:151–168.

Reuter, H., Böckmann, S., Breckling, B. (2008) Analysing Cross-pollination studies in maize. In: Breckling B, Reuter H, Verhoeven R (eds.): Implications of GM Crop Cultivation at Large Spatial Scales. Peter Lang 14, Frankfurt, 47–52.

Schmidt, G., Kleppin, L., Schröder, W., Breckling, B., Reuter, H., Eschenbach, C., Windhorst, W., Höltl, K., Wurbs, A., Barkmann, J., Marggraf, R., Thiel, M. (2009): Systemic Risks of Genetically Modified Organisms in Crop Production: Interdisciplinary Perspective. Gaia 18(2):119–126.

Schmidt, G., Schröder, W. (2011a) Implications of GMO cultivation and monitoring-series – Editorial. Environmental Sciences Europe 23(2), doi: 10.1186/2190-4715-23-2

Stoppe-Ramadan, S., Winter, G. (2010): EU and German law on coexistence: Individual and systemic solutions and their compatibility with property rights. In: Breckling B, Verhoeven R (eds.): Large-area effects of GM-crop cultivation. Theorie in der Ökologie 16. Peter Lang, Frankfurt, 21–124.

Struss, J., Winter, G. (2012) Rechtliche Regelung systemischer Risiken von GVO: (Ökologische Modellierung und ihre juristische Verwertbarkeit. In: Breckling B, Schmidt G, Schröder W (Hrsg.): GeneRisk – Systemische Risiken der Gentechnik: Analyse von Umweltwirkungen gentechnisch veränderter Organismen in der Landwirtschaft. Springer, Heidelberg, 151–162.

Williams, S. (2002) Free as in freedom. Richard Stallman's crusade for free software. Cambridge, O´Reilly.

Links

[1] http://www.springer.com/life+sciences/biochemistry+%26+biophysics/book/978-3-642-23432-3
[2] http://www.enveurope.com/series/GMO_cultivation
[3] http://www.isaaa.org/resources/publications/briefs/42/executivesummary/default.asp
[4] http://www.sozial-oekologische-forschung.org/de/692.php
[5] http://www.mapserver.uni-vechta.de/generisk/generisk_info_portal/index.php?webgis=1
[6] http://inspire.jrc.ec.europa.eu
[7] http://www.geoportal.de
[8] http://www.mapbender.org
[9] http://sia.eionet.europa.eu/CLC2000
[10] http://apps2.bvl.bund.de/stareg_web/showflaechen.do

Chapter III

Long-term experience and sociological consequences

Impacts of genetically engineered crops on pesticide use in the U.S. – The first sixteen years

Charles Benbrook

Center for Sustaining Agriculture and Natural Resources, Washington State University, Pullman, Washington, USA

Extended Abstract[1]

Herbicide-tolerant (HT) corn, soybean, and cotton cultivars, and Bt-transgenic corn and cotton, have been remarkable commercial successes in the United States since their introduction in 1996. Claims are often made that these technologies have, and continue to reduce pesticide use. Annual corn, soybean, and cotton pesticide use data collected by the U.S. Department of Agriculture (USDA) constitutes the most complete public dataset with which to assess the impacts of GE crops on the kilograms of pesticides applied.

A model was developed to quantify by year the pesticide use impacts per hectare planted of the six major commercial GE pest-management traits: HT corn, soybeans, and cotton; Bt corn for control of the European corn borer (ECB), Bt corn for the corn rootworm (CRW); and Bt cotton for Lepidopteron (budworm/bollworm complex) insect control. Changes in pesticide applications brought about by the planting of a GE-trait hectare were estimated by crop, year and trait, and aggregated across all GE trait hectares planted over the 16 year period 1996–2011.

HT crop technology has led to ~ 239 million kilogram increase in herbicide use across the three major GE-HT crops, while Bt corn and cotton has reduced insecticide applications by 56 million kilograms. The reduction in insecticide use associated with Bt corn and cotton has, however, been accompanied by the biosynthesis of substantial volumes of Bt Cry endotoxins, ranging from 0.5 kgs/hectare to 4.2 kgs/hecatre in the case of SmartStax Bt corn. Overall, pesticide use increased by an estimated 183 million kilograms, or about 7 %, between 1996 and 2011.

Looking ahead, the spread of glyphosate-resistant weeds in HT weed-management systems is bound to trigger further increases in the intensity of herbicide use. The volume of 2,4-D sprayed on corn, e.g., could increase 15-fold by 2019 from 2010 levels

---

[1] A full paper is published in Environmental Sciences Europe, Series: Implications of GMO-cultivation and monitoring: Benbrook, C. (2012) Impacts of genetically engineered crops on pesticide use in the U.S. – the first sixteen years. ESEU 24:24.)

if the USDA approves unrestricted planting of 2,4-D HT corn, a technology that could add 2.1 kg additional herbicide/ha, about a 50% increase. Further increases are expected in the use of 2,4-D and dicamba on newly HT corn, soybean and cotton varieties, and the volume applied of other herbicides linked to newly-HT crop cultivars will surely rise. The increase in herbicides applied on HT hectares has dwarfed the reduction in insecticide use over the 16 years, and will continue to do so for the foreseeable future.

Breckling, B. & Verhoeven, R. (2013) GM-Crop Cultivation – Ecological Effects on a Landscape Scale. Theorie in der Ökologie 17. Frankfurt, Peter Lang.

Farmer's choice of seeds in five regions under different levels of seed market concentration and GM crop adoption

Rosa Binimelis[1], Angelika Hilbeck[2], Tamara Lebrecht[2], Rapahela Vogel[2] & Jack A. Heinemann[3]

[1] Center for Agro-food Economy and Development-CREDA-UPC-IRTA, Castelldefels, Spain; [2] Swiss Federal Institute of Technology, Institute of Integrative Biology, Zurich, Switzerland; [3] Centre for Integrated Research and School of Biological Sciences, University of Canterbury, Christchurch, New Zealand

Extended Abstract[1]

The issue of access to seeds/cultivars is important to genetic resource preservation and future food security. However, market concentration of staple food seeds is occurring in all sectors of the food production chain "from farm to fork". It has been well documented and has lead to accelerated the concentration of the seed market and the shift of breeding efforts from the public to the private/industrial domain. Genetic engineering of crop plants has been suggested to be a key technology for future crop plant enhancement and accelerated breeding. However, the technology continues to be highly controversial in many countries and, in particular, in Europe finds little acceptance with consumers. Hence, only one GM crop, Bt maize Mon810, is commercially grown at a large scale in one country only, Spain. Moreover, it has been repeatedly suggested that countries that do not adopt GM crops do or will have fewer options. Or, vice versa, that countries that have so far rejected GM crops have had an impact on their productivity. We found it timely to investigate these repeated claims and wanted to know how cultivar choices have developed in GM-adopting vs non-adopting countries in Europe.

We used surveys of seed catalogues from local and regional seed suppliers, transnational seed corporations and public national and European seed registration catalogues as an approximation for real world choices available to farmers, to estimate how much choice maize farmers have in four countries (Austria, Germany, Spain, Switzerland), with different degrees of GM crop adoption. We analysed how the availability of maize cultivars for farmers in these countries has developed since the mid 1990ies when GM maize was introduced to agriculture globally. Non-adopting countries were Austria, Germany, and Switzerland and the GM-adopting country was Spain.

---

1 A full paper is submitted to Environmental Sciences Europe, Series: Implications of GMO-cultivation and monitoring.

We found no evidence that restrictions and regulations of GM crops in Europe have led to less choices for farmers in the non-adopting countries Austria, Germany and Switzerland. Quite in contrast, we observed that in Spain, which has adopted GM maize, the seed market was more concentrated with fewer differentiated cultivars on offer for farmers. In Spain, overall numbers of maize cultivars decreased because an increasing number of non-GM cultivars is being replaced by a lower number of GM cultivars. In stark contrast, in Germany, maize cultivar numbers almost tripled from 116 varieties in 1994 to around 320 varieties in 2011, the steepest increase among the non-adopting countries. Hence, we concluded that in non-adopting European countries, farmers have more maize cultivars available to them today than they had in the 1990s despite restricting GM-varieties while farmers in GM-adopting countries have less cultivars available.

In the United States, where farmers have primarily a choice among GM-varieties for maize and soybean, this lack of choice of non-GM varieties is causing increasing concerns and motivating various grass-roots organizations to begin local breeding efforts. Adaption to the rapidly changing climate and increased need for ecological intensification will fundamentally hinge upon free accessability of farmers and breeders to plentiful seeds.

Implications of GM crops in subsistence-based agricultural systems in Africa

Denis W. Aheto[1], Thomas Bøhn[2,6], Broder Breckling[3], Johnnie van den Berg[4], Lim Li Ching[5] & Odd-Gunnar Wikmark[2]

[1] School of Biological Sciences, University of Cape Coast, Ghana; [2] GenØk – Centre for Biosafety, Tromsø, Norway; [3] University of Vechta, Germany; [4] Unit for Environmental Sciences and Management, North-West University, South Africa; [5] Third World Network (TWN), Malaysia; [6] University of Tromsø, Norway

Abstract

Africa has deep contentions on the use of GM crops in agriculture, similar to those found in Europe and elsewhere. However, it is apparent that the debate is most protracted on the continent with two entrenched viewpoints i.e. the pro-GMO and anti-GMO groups. The challenge for an acceptable consensus is attributable to a complexity of issues relative to the introduction of GM maize into small-scale farming systems that fundamentally relies on open pollinated varieties (OPVs) with broad genetic backgrounds and tolerance to diverse biotic stresses, and which is usually produced for the informal seed market. Other factors relate to the generally low capacity of African states and weak mechanisms for assessing the potential risks posed by GM crops. The lack of public awareness, participation and information sharing are additional limiting factors.

These issues have weakened government and policy responses to the potential deployment of GM crops on the continent. This review draws on research-based evidence as a basis to comment on some key issues to inform the development of biosafety standards in African countries. We conclude that the potential introduction of GM crops into small-scale farming would lead to huge consequences from emerging ecological, economic and trade impacts if these issues raised are not taken into account in decision-making processes.

Introduction

The objective of this review is to draw attention to some key issues within the African context relevant for improving biosafety implementation efforts on the continent. The task of predicting how the presence of transgenes in agricultural crops is likely to influence the ecology and development of a recipient environment and society is highly demanding and requires best available knowledge from a number of different disciplines (Myhr & Traavik 2007). Moreover, socially responsible actions must also be based on knowledge about the cultural context and agricultural practices, and the level and demands of farmers and the seed industry (Mugo et al. 2005). The arguments in favour of genetically modified (GM) crops for small-scale farmers center on selective advantage in the form of insect pest resistance or herbicide tolerance with the promise of higher productivity. However, both insecticidal traits (Bt) and herbicide tolerance traits (HT) raise concern for long-term reliability against resistance development, biodiversity conservation, food security and environmental sustainability in African agriculture and related ecosystems. The perceived benefits of growing GM crops for poverty alleviation in Africa must be evaluated together with possible conflicts posed to the environment as well as with the bigger picture of food insecurity on the continent (deGrassi 2003).

Environmental uncertainties related to GM crops, and in particular for Bt maize include potential development of resistance among target insects, non-target adverse effects on beneficial organisms and cross hybridization with non-GM varieties, with subsequent loss of biological and genetic diversity (Andow & Zwahlen 2006). The contention on the use of GM crops in small-scale farming has been on this basis. Presently, the debate has assumed another dimension. The argument stands to reason that whether GM crops are grown or not within small-scale systems, one must take into account that the African context has a low level of control, testing, monitoring and possible response measures. Furthermore, whether modern biotechnology will meet the needs of poor producers and whether the capability exists in current input markets to deliver the seed and information that embodies GM technology have long been debated (Tripp 2001). This makes Africa at large particularly vulnerable to potential unintended and undesirable spread of GMOs due to a consecutive mixing with non-GM material. Maize seed is easy to store and transport and through pollen flow, characteristics can easily be transferred between varieties (Smale & De Groote 2003). Chances are high that GM crops, as well as their associated food and feed products traded on the world market could still find their way into countries and places where they were originally not anticipated or accepted. Transgenes could spread across national borders regardless of whatever policy exists. However, it has been indicated that Africa needs to strengthen its capacity both with regards to research and development as well as with regard to legal and policy aspects of biosafety (Eicher et al. 2006). This review aims at discussing implications of GM crop introductions in African food systems in particular maize based on science-based evidence. Possible impacts are discussed and we suggest ways forward for how the challenges could be mitigated or resolved, with the intention to improve the African biosafety situation.

Agricultural structure

The agricultural structure in most of sub-Saharan Africa is not only small-scale but typically dense. Dominance of small fields with relatively few larger fields in the neighbourhood is common. This type of agricultural setup would facilitate the possibility of transgene flow through higher cross-pollination among small field neighbours

(Aheto et al. 2011). Maize has a high risk of gene flow through cross-pollination, particularly when landholdings are fragmented, varieties are planted contiguously, and farmers recycle, exchange, or mix maize seeds (Smale & De Groote 2003). This is of special interest to estimate impacts not only on smallholders but also on a wider number of smaller fields with potentially diverse locally-grown seeds. GMOs if introduced would clearly spread and diffuse transgenes due to high rates of cross-pollination among neighbouring fields (Aheto 2009; Bøhn et al., this issue). This would pose a major obstacle in maintaining GMO-free zones as proposed in the African Model Law on Safety in Biotechnology. GM crops, if introduced, would pose a major challenge to maintain GMO-free zones also due to the prevailing systems of seed acquisition and local exchange which would pose a further complexity. High density of fields poses a major difficulty to practice any legal isolation distance requirements within the small-scale setting. In this context, the presence of feral maize on the landscape would contribute to the unintended persistence of transgenes in the local maize gene pool.

Farming within the African context is mostly operated on small plots of land mostly in a size range of below two hectares (Fig. 1). Seed saving guarantees multi-year seed supply among farmers as an important cultural practice that enhances local seed diversity, crop improvement and secures household food security (deGrassi 2003). For example, in smallholder agriculture in Kenya, open pollinated varieties and seed-saving and exchange are common (Mwangi & Ely 2001). Similar practices are implemented throughout Africa (Smale & Phiri 1998; Smale & DeGroote 2003) and also in South Africa where there is a strong and regulated private seed sector, seed production and marketing system (Mphinyane & Terblanché 2005). The use of genetically modified varieties would limit options for these traditional practices and put farmers and their households at risk. The geometry of fields, mostly small, dense and often close to each other renders gene flow highly likely (Tab. 1). These issues have adverse implications for organic and conventional food production but also for the purity of non-GM seed production.

Table 1: Features of small-scale maize farming within the African context.

Feature	Ghana	Zambia
Settlement description	Urban periphery (Accra, 2006)	Rural area (Chongwe, 2012)
Area of investigation	1 km^2	1 km^2
Number of fields per km^2	58	97
Average field acreage from GIS records (ha) (minimum – maximum acreage in m^2)	0.81 (1.0 – 35000.0)	0.49 (13.0 - 43609.0)
Fractional maize area as a % of total maize acreage	4.5	48.0
Range of distances to next field (m)	5 – 10	2 – 10

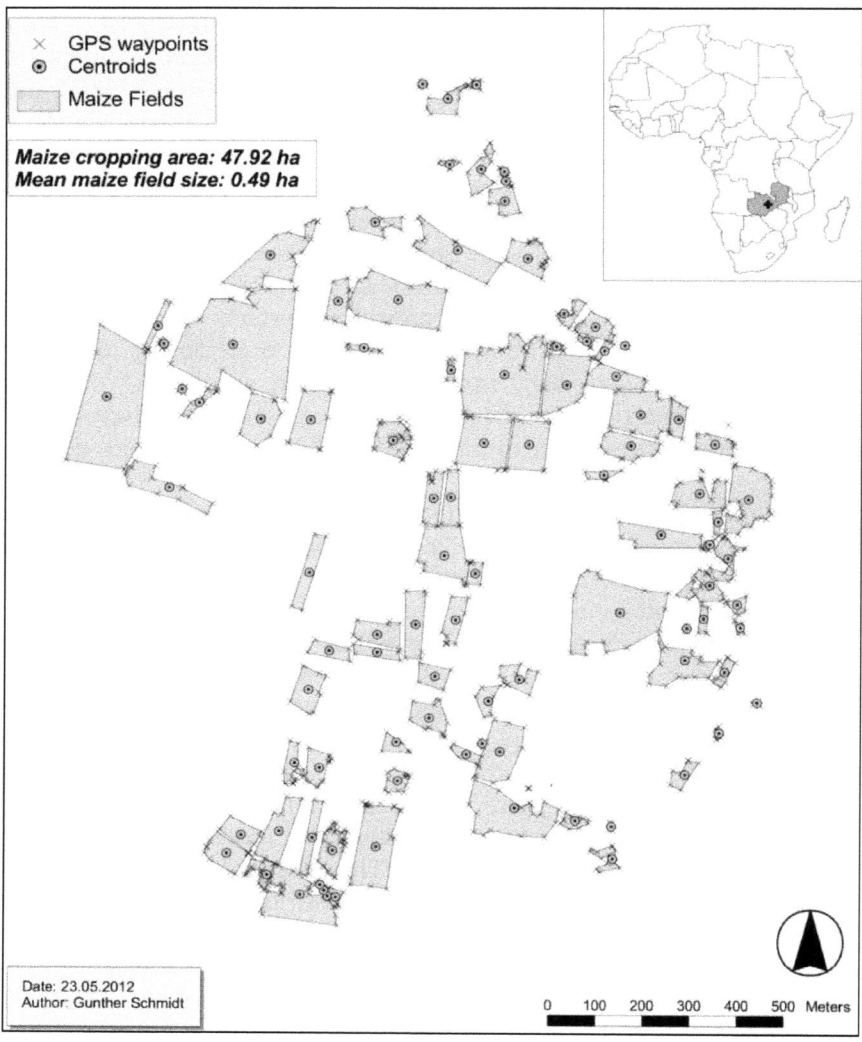

Figure 1: Example of a typical small-scale maize agriculture in Africa encompassing a large sector of smallholder farming crucial for food security (Chongwe Province in Zambia, March 2012).

Administrative and regulatory competencies

Establishing the capacities to monitor the fate of transgenic DNA ("transgenes") in the environment is an essential element in any biosafety effort. Any regulation of genetically modified organisms (GMOs), including market-driven differentiation and governmental regulation of transgene-containing products must include some measure of detection and monitoring of transgenes. On the other hand, transgenes also offer unique markers with which to trace the flow of genetic materials in general through complex ecological situations. Genetically modified crops are traded on the world market and therefore require countries to set up regulations for transboundary movements. The

Cartagena Protocol on Biosafety (2000) is the only international treaty specifically regulating GMOs and all parties have to take legal, administrative and other measures to implement the protocol, which often includes the development of national biosafety regulations. In the African context, regulations as well as enforcement on biosafety are still largely limited. The UNEP-GEF and the African Union operated framework Projects and now NEPAD have made modest gains in assigning some administrative competencies. Unfortunately independent risk assessment data that draws on regionally acquired environmental data is still widely lacking (Aheto et al. 2011). For biosafety, this is problematic since GM varieties on the world market are continuously being developed and notified for conditions in Africa that differ in climate, agricultural structure, interacting organisms (both pests and beneficial organisms) and differ in consumer preferences. There are therefore highly relevant gaps in this field required to be filled.

Seed use and seed exchange practices

A majority of traditional farmers acquire seeds for planting from a wide variety of sources within the informal sector. Acquisition of seeds as gifts from neighbours or home-saved from previous harvests are relevant sources. Commercial procurement of seeds does not rely upon a need for insect resistant varieties but rather on more stable high yielding varieties that may be shared among farmers in subsequent seasons (Fig. 2). Farmers would like to grow different crop varieties, i.e. land races. Also, seed exchange among farmers limits the possibility of co-existence of farming systems involving conventional and GM crop farming, possibly lowering the economic value for conventional and organic food producers and causing a decline in crop genetic purity.

A crucial factor relates to the non-distinction between food grain and seed grain by small farmers. Therefore any GM food import or aid would eventually end up in the cultivation systems of farmers. Another critical issue also relates to the fact that commercially-oriented subsistence farmers, in many cases, procure seeds from formal seed stores with a notion to benefit from high-yielding and early maturing varieties. The trait of insect-resistance is not among the most important options when it comes to choice of seeds for planting, especially if the target pest is a minor problem and farmers cannot physically observe the protection provided by the Bt trait (Assefa & van den Berg 2010). Most often, agronomic factors such as yield potential, drought tolerance, husk cover of ears, as well as resistance to storage pests and rain damage may be overriding factors in hybrid preference amongst farmers (Assefa & van den Berg 2010; Grouse et al. in review). Farmers may not discern the benefits from the inserted trait, or may view these as less important than some other disadvantageous traits of the new variety relative to those they currently grow (Smale & De Groote 2003).

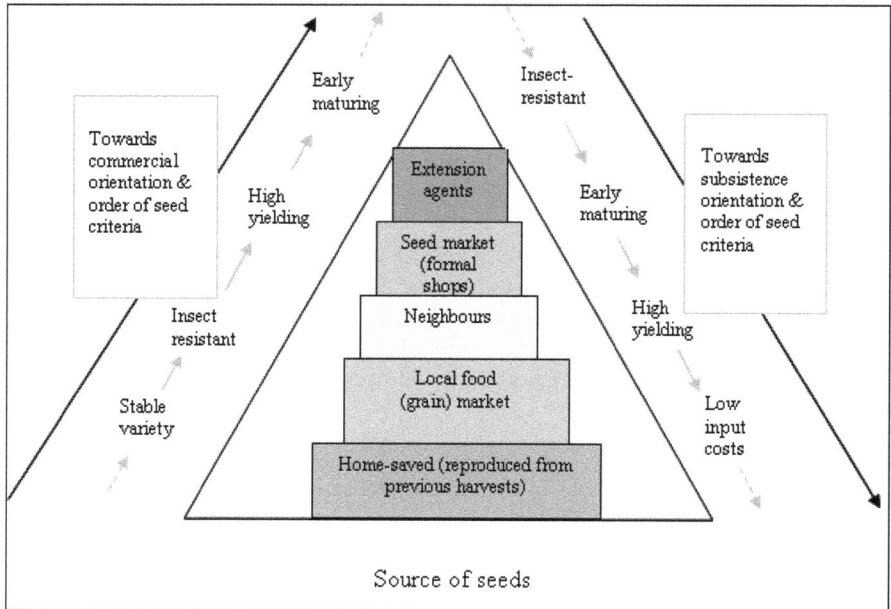

Figure 2: Conceptual model of farmer seed selection criteria (modified from Aheto 2009).

Coexistence and trait segregation

Systems of seed use and exchange will impose further complexity to coexistence between GM and non-GM since distinction is hardly made between food grain and seed grain with respect to seed sown, as smallholders cultivate various landrace open-pollinated varieties alongside available commercial hybrids. This makes trait segregation difficult if not impossible. Therefore, coexistence of GM and local non-GM systems as being tested under the European conditions would be impractical for the African situation. In theory, different planting times among neighbouring farmers seem not practically implementable because most small-scale farmers do first plantings to take advantage of the rains when it comes, i.e. not under their control. Thus, co-existence of GM and non-GM crops would be an impractical scenario under the small-scale context.

Food security, genetic resources conservation and trade

A high number of African households are supported per acre of farm with women constituting the majority. Therefore, should GM crops/foods be adopted, these should be under conditions that avoid potential risks. Time and effort must be devoted to on-farm trials before any interventions in this regard. Policy makers and researchers in developing countries should carefully assess environmental and socioeconomic risks (such as the major risks to biodiversity, the prospects of insufficient out-crossing distances, the relative absence of clear labeling and other threats to seed purity from adjacent traditional food production) before farmers change their conventional farming methods to GM (Azadi & Ho 2010). Introduction of GMOs could lead to transgenic contamination of local seeds, with a high probability of impacting areas where organic food farming is likely practiced. This could potentially lead to economic and social impacts. Consequences on native biodiversity cannot be ruled out. Such infringement of natural borders has been described as a crucial ethical concern. Also, options for improving conventional or organic export trade and foreign income will be closed under an agricultural system where GM and non-GM crops are mixed up. The documentation of crop purity is a must for premium price products in the organic trade, a niche in growth and in developed and developing countries (Hewlett & Azeez 2007). Crop management traditions, including local adaptation to food security through seed saving cultures of landraces must therefore be upheld and protected.

Consumer choice, liability and patent infringement

The possibility to allow for consumer choice and trait segregation is largely minimal. Containment of GM products, including potential mitigation or removal from the environment if undesirable results should become apparent would require serious and realistic provisions to enforce within the African context. Central planning of production and regulation of agricultural products marketing would be difficult to implement. To orient producers to meet the needs of consumers would constitute a daunting task since the producers are mostly the consumers and vice versa. Since transgenes are protected by national and international laws as intellectual property (IP) (Heinemann 2007), small farmers who unintentionally grow GM crops in their farms may face legal actions. Patent infringement is considered a very serious issue and even the use of certain transgenes and locally adapted GM crop varieties may be problematic (Mugo et al. 2005). The case of Percy Schmeiser versus Monsanto provides clear indications that infringement of IP may be followed up even in a situation where the farmer does not know or does not want the transgenic trait. This could be a serious issue for example in organic production. Patent infringement under the provisions of the Canadian Patent Act for example was seen as illegal even when farmers were inadvertently contaminated by neighbours who cultivated GM crops (Heinemann 2007). Should a similar case occur in the African context, it might negate small farmers' rights to save and replant their own seeds and thus break down long histories of traditional culture.

Monitoring implementation

Biosafety measures are comparatively more difficult to implement within the African context, in relation to the developed countries, since the resources that can be invested in the establishment of anticipatory regulatory efforts – including monitoring and enforcement – are substantially less. On a policy level, some gains have been achieved through the established National Biosafety Frameworks (Mclean et al. 2002), which provide basic regulatory guidance regimes, as a basis to move forward. As indicated by Eicher et al. (2006) African governments need to develop a national biotechnology strategy that defines how biotechnology fits into the overall national agricultural research strategy, agricultural development strategy and target farmers and sectors where biotechnology tools will be applied based on needs and priorities identified by various stakeholders. A key measure would be to effectively regulate GM food and feed products imported into the country since once they are admitted; it is not feasible to control their further spread and diffusion into on-farm seed stocks and into the environment.

Public participation

Maize has a critical role for nutrition across the African continent. It is the most important staple food crop. Therefore, public awareness on GMO issues and discussion on implications must be enhanced (Egziabher 2007). People have the right to participate and contribute to decision-making when it is about their staple crop, grown in gardens and fields, guided by their own traditional knowledge and culture. Coherence in the administrative and regulatory structures is limited, as could be seen in the operation of extension services in urban areas. Dealing with uncertainties and contradictions are among key areas that need to be addressed. Efficient programmes should be developed to support farmers to improve their responses towards effective seed saving and cropping management. It should however be acknowledged that to orient smallholder producers to meet the demands of consumers will be difficult to achieve in Africa (Mugo et al. 2005) because producers are also the consumers. The most important freedom of choice for farmers in Africa seems to be the ability and right to save and share seeds without risk of contamination by transgenes and to maintain seed availability and exchange as an open system.

Research financing on risk assessment

A major setback is the limitation in the capacity and financing of biosafety initiatives. This affects biosafety implementation efforts and limits the effectiveness of risk assessment procedures. Independent risk assessment research that is not influenced by external business interests is a requirement for a credible regulation and administration acting in the interest of the general public. Risk assessment is the identification and evaluation of potential adverse effects of genetically modified organisms on the conservation and sustainable use of biological diversity in the potential receiving environment, taking into account also risks to human health (Cartagena Protocol 2000).

African countries generally have limited scientific competence to monitor, research and conduct risk assessments that examine the full health, environmental and socio-economic implications of genetic engineering and GMOs. For African countries, the entering into force of the Cartagena Protocol on Biosafety in 2000 provided an important benchmark for addressing risk-related issues of GMOs. The Protocol stipulates that the national approval of a GMO should be based on prior informed consent. For approval, a complete risk analysis is required. If the variety was developed in an African country, the effort has to be made completely in that country. If it is a foreign variety for which an applicant seeks consent, the risk evaluation should take into account previous risk studies but also complement it with additional information that covers specific conditions of the country for which consent is being sought (Aheto 2009).

Risk assessment therefore is expected to cover the full spectrum of relevant effects i.e. direct or indirect, immediate or delayed, cumulative and detrimental effects of GMO on biodiversity, environment and human health (Reuter et al. 2008). In the event of an irreversible or detrimental effect, decisions should be based on the precautionary principle. If the effects are assumed to be negligible, they will be considered irrelevant and ignored. If effects are found to be relevant or significant, then the assessment of risk should be extended and broadened. However, it is important to note that risks that are irrelevant in one environment may not be irrelevant in another. Thus, in risk analysis of GMO, a combined knowledge on scientific and technical procedures is necessary. Capacity building is crucial for the implementation of effective biosafety standards on the African continent. The majority of African countries lack capacity to execute effective biosafety investigations including laboratory testing according to the requirements of their own environmental and social situation.

Conclusions

Large-scale consequences of small-scale farming with GM crops are implied in this analysis. The introduction of new technologies that do not take into account the contextual factors in Africa will cause problems and ultimately fail. From an ecological perspective, introduction of GM crops would lead to uncontrolled large-scale spread and persistence of transgenes within the small-scale agricultural systems in Africa with unpredictable recombination and evolution in crop meta-population. The socio-cultural implications relate to intellectual property rights, which threaten traditional seed use patterns. Impurities in harvest would prevent development and export-options. Major challenges in regulatory decision-making are envisaged since traceability, administrative regulation and resistance management regimes are difficult to impossible. Furthermore, huge increases in administrative costs are expected owing to laboratory analysis, monitoring and assuring quality control in testing.

This paper therefore makes a strong call for a precautionary approach to biosafety in the face of uncertainty. Independent research and biosafety capacity building across Africa are top-ranked priorities in food security. The question of transgene flow in crop plants must be addressed as a meta-population problem, since transgenic plants will be exchanged between local (and over time more remote) seed pools due to human agency, triggered by individual farmer decision-making, commercial forces, governmental regulation, etc. The convergence of ecological, commercial, cultural and regulatory interests in understanding flows of transgenes is perhaps most noticeable in the agricultural environment of major crops.

Acknowledgements: Funding is acknowledged from Genøk – Center for Biosafety, Tromsø, Norway for funding field studies in Zambia as part of compilation of this work and presentation at the Third International Conference on Large-Scale Implications of GMO Cultivation at Large Spatial Scales in Bremen, Germany in June 2012. The authors are highly indebted to the many farmers who continue to cooperate with us to do research on these issues.

References

African Draft Model Law on Safety in Biotechnology (2001).
http://www.africabio.com/policies/MODEL%20LAW%20ON%20BIOSAFETY_ff.htm

Aheto, D.W., Reuter, H., & Breckling, B. (2011) A modeling assessment of geneflow in smallholder agriculture in West Africa. Environmental Sciences Europe 2011 23:9. SpringerOpen DOI: 10.1186/2190-4715-23-9

Aheto, D.W. (2009) Implication Analysis for Biotechnology Regulation and Management in Africa. Baseline Studies for Assessment of Potential Effects of Genetically Modified Maize (Zea mais L.) Cultivation in Ghanaian Agriculture. Frankfurt Am Main, Theorie in der Ökologie, Peter Lang, Vol. 15, 240pp.

Andow, D.A. & Zwahlen, C. (2006) Assessing environmental risks of transgenic plants. Ecology Letters: 9: 196–214.

Assefa, Y., van den Berg, J. (2010) Genetically modified maize: Adoption practices of small-scale farmers in South Africa and implications for resource-poort farmers on the continent. Aspects of Applied Biology 96: 215–223.

Azadi, H., Ho, P. (2010) Genetically modified and organic crops in developing countries: A review of options for food security. Biotechnology Advances 28: 160–168

Cartagena Protocol on Biosafety (2000) Convention on Biological Diversity, CBD http://www.biodiv.org/biosafety/protocol.shtml, accessed on 4.05.07.

deGrassi, A. (2003) Genetically Modified Crops and Sustainable Poverty Alleviation in Sub-Saharan Africa; An assessment of current evidence, Third World Network-Africa Publisher Website: www.twnafrica.org, accessed 18.6.2008

Eicher, C.K., Maredia, K., Sithole-Niang, I. (2006) Crop biotechnology and the African farmer. Food Policy 31: 504–527.

Egziabher, T.B.G. (2007) The Cartagena Protocol on Biosafety: History, Content and Implementation from a Developing Country Perspective In: Traavik, T. & Li Ching, L. (eds.) Biosafety first – Holistic approaches to risk and uncertainty in genetic engineering and genetically modified organisms. Tapir academic press. 389–405.

Gouse, M., Pray, C., Schimmelpfenning, D., Kirsten, J. 2006. Three seasons of subsistence insect-resistant maize in South Africa: have small farmers benefitted? AgBioforum 9: 15–22.

Heinemann, J. A. (2007) A typology of the effects of trans(gene) flow on the conservation and sustainable use of genetic resources. Rome UN FAO: 1–94.

Hewlett, K. & Azeez, G. (2007) The Economic Impacts of GM Contamination Incidents on the Organic Sector In: Proceedings of the Third International Conference on Coexistence between Genetically Modified (GM) and non-GM based Agricultural Supply Chains, Seville (Spain), 20/21 November 2007. 336–337.

Mclean, M.A., Frederick, R.J. Traynor, P.L., Cohen, J.I., Komen, J. (2002) A conceptual framework for implementing biosafety: Linking policy, capacity, and regulation. ISNAR Briefing Paper 47. ISNAR, The Hague, Netherlands.

Mphinyane, M.S., Terblanché, S.E. (2005) Personal and socio-economical variables affecting the adoptin of maize production intervention program by dryland farmers in the Vuwani district, Limpopo province. South African Journal of Agricultural Extention 35: 221–240.

Mugo, S., De Groote, H. Bergvinson, D., Mulaa, M., Songa, J., Gichuki, S. (2005) Developing Bt maize for resource-poor farmers – recent advances in the IRMA project. African Journal of Biotechnology, 4: 1490–1504.

Mwangi, P.N., Ely, A. (2001) Assessing risks and benefits: Bt maize in Kenya. Biotechnology and Development Monitor 48: 6–9.

Myhr, A.I., Traavik, T. (2007) GE Applications and GMO release: The ethical challenges In: Traavik, T. & Li Ching, L. (Eds) Biosafety First. Trondheim. Tapir Academic Press. 123–135.

Reuter, H., Breckling, B., Böckmann, S. (2008) Modelling maize hybridisation on a landscape level. Data analysis and development of a dispersal kernel. Proceedings, GMLS 2008, Bremen (www.gmls.eu).

Schmeiser, P. (1999) Monsanto vrs. Schmeiser. The Classic David and Goliath Struggle. http://www.percyschmeiser.com

Smale, M., DeGroote, H. (2003) Diagnostic research to enable adoption of transgenic crop varieties by smallholder farmers in Sub-Saharan Africa. African Journal of Biotechnology 2: 586–595.

Smale, M., Phiri, A. (1998) Institutional change and discontinuities in farmers' use of hybrid maize seed and fertilizer in Malawi: Findings from the 1996-7 CIMMYT/MoALD Survey. Economics Working Paper 98-01. International Maize and Wheat Improvement Center (CIMMYT), Mexico City.

Tripp, R. (2001) Can biotechnology reach the poor? The adequacy of information and seed delivery. Food Policy, 26: 249–264.

Co-existence challenges in small-scale farming when farmers share and save seeds

Thomas Bøhn[1,5], Denis W. Aheto[2], Felix S. Mwangala[3], Inger Louise Bones[1], Christopher Simoloka[3], Ireen Mbeule[3], Odd-Gunnar Wikmark[1], Gunther Schmidt[6] & Ignacio Chapela[1,4]

[1] GenØk – Centre for Biosafety, Tromsø, Norway; [2] School of Biological Sciences, University of Cape Coast, Ghana; [3] National Institute for Scientific and Industrial Research, Zambia; [4] University of California, Berkeley, US; [5] University of Tromsø, Norway; [6] University of Vechta, Germany

Abstract

Gene flow by means of pollen and seeds in maize influences local, regional and global maize biodiversity. Developing countries are centers of diversity for maize and preserve seeds also in informal seed systems. Particularly in poor communities, seed saving and sharing often co-occur with farming on small fields. We present preliminary investigations from a small-scale maize farming community, in Chongwe, Zambia, to illustrate the significance of seed saving and sharing for patterns of gene flow. The potential introduction of genetically modified (GM) plants brings in new dimensions of challenges for farmers e.g. related to: i) co-existence of GM and non-GM varieties; ii) potential infringement of intellectual property rights; and iii) trans-boundary movement of products to countries that do not accept certain GM products. Small-scale farming is vulnerable to cross-contamination due to limited separation between fields. If transgenes are introduced into small-scale agricultural contexts, uncontrolled diffusion and further spread seems unavoidable. Removal of transgenes as well as the regulatory implications of transgenes would require control of innumerable small informal seed stores kept by farmers.

Introduction

Maize is an important cereal crop and staple food in many regions of the world, especially in Sub-Saharan Africa. In this study we provide concepts necessary for the elucidation of gene flow, with emphasis on potential transgene flow, in an African environment where small-scale and subsistence farms are dominant landscape components. We identify basic needs for measurement and analysis, and chart future needs for research in this field. It is worthy to note that from its center of origin, in Mesoamerica, farmers have for thousands of years experimented, spread, mixed and selected favorable maize plants for different environmental conditions on all continents. This diversification process has until relatively recent times been based on informal seed systems, free sharing and no intellectual property rights on genes or germplasm. The ultimate consequence of this primarily seed-based maize gene-flow experiment, driven by farmers, is the highly appreciated richness of maize food and culture that we have today.

However, gene flow studies in maize have focused on the role of pollen movement and cross-hybridization (Dyer et al. 2009; Gray et al. 2011). Maize cross-pollination decreases rapidly within 30 m, but includes a long tail with low cross-pollination occurring over several hundred meters (Beckie & Hall 2008). Cross-pollination is still not zero at 800 m distance (Kawashima et al. 2011). Open pollinated varieties have higher outcrossing than hybrid cultivars (Sanvido et al. 2008).

Pollen flow has become particularly relevant with the introduction of transgenic or genetically modified (GM) plants which are protected by intellectual property rights (IPR) and bringing with them specific regulatory and risk assessment demands. Specific recommendations for isolation distances have been proposed to ensure that products comply with the EU labelling threshold of 0.9%. For small-scale farming, proposed isolation distances are varying, e.g. 20 m (Gustafson et al. 2006), or 25–50 m (Beckie & Hall 2008). Further, a pollen barrier of 20 m has been recommended (Messeguer et al. 2006). Devos et al. (2005) argued that large fields (>5 ha), do not need any isolation distance but small fields (<1 ha) requires a 50 m isolation distance. In a South-African context, (Viljoen & Chetty 2011) recommended isolation distances of 135 m for cross-pollination levels between <1.0% and 0.1%, 503 m for <0.1% to 0.01% and 1.8 km for <0.01% to 0.001% based on high cross–pollination values. In Zambia, isolation distances recommended in order to ensure seed purity in the production of certified maize seed and basic maize seed is 200 m and 400 m, respectively (MAFF 1995).

In the context of small-scale farming, quantitative recommendations as referred to above may clearly be of little or no use when i) fields are smaller than the isolation distance between fields, and ii) isolation barriers would consume a significant part of individual fields.

Adventitious presence of transgenes may also come from the spread of seeds. Seed sharing make people effective vectors of genetic material; over large distances, over natural and regulatory borders. This is difficult to manage and control, even in Europe where tighter controls may be expected than in developing countries. Bannert and Stamp (2007) reported from a GM maize field trial in Switzerland that unwanted mixing of seeds may contribute more to the GM content in receptor fields than the pollen for cross-pollination.

In this study we present preliminary data on field sizes and spatial arrangement of fields, as well as data from questionnaires about farmer activities like seed saving and sharing, in a small-scale maize farming community in Chongwe, Zambia. We discuss the link between gene flow and the feasibility for co-existence of GM maize (if introduced) and non-GM maize in this particular agricultural context.

Methods

We mapped the spatial patterns of cultivated fields in a maize farming community of Chongwe, Zambia, using GPS devices (Garmin 62S) in March 2012. The coordinates (UTM, GCS Arc 1950) of the waypoints of each field were transformed to a polygon layer by means of ArcGIS 10.0, and field centroids and acreage were calculated. Analysis of the spatial distribution, and average distances to nearest field neighbours with respect to field centroids were performed using SIMULA-based written programs (Dahl 1968). Additionally, we interviewed farmers (n = 32) that used these same fields (but not all of them) about their practice related to seed saving and sharing.

Results

A total of ninety-seven fields were found within an area of 1 km^2 in the Zambian community of Chongwe. The mapped fields were small with an average size of about half a hectare (0.49 ha). The fractional area of cultivated maize as a percentage of the total study area was 48%. Two thirds of the fields were smaller than 0.5 ha and only 5% of the fields were larger than 2.0 ha. All fields had neighbouring fields that were closer than 10 m away. Based on data from questionnaires, about two thirds of the farmers both saved and shared seeds from the previous season, involving both local varieties and commercial hybrids. Sharing happened mostly within the local community (i.e. to very nearby farmers and up to 800 m away), but also across communities, i.e. up to a distance of 100 km.

Discussion

Much progress has been made in understanding transgene flow in large, industrial agricultural settings, but by contrast our knowledge of gene flow is very restricted for cases where small-scale and subsistence agriculture dominates the ecosystem. In particular, there is a knowledge gap on this issue in the case of Africa.

The community of Chongwe, Zambia, represents a small-scale farming system in Africa. Given the density, sizes and distances between fields we observed, pollen will interchange genetic material between farmers at a high rate. Thus, this kind of farming seems incompatible with a segregated and parallel growing of GM and non-GM maize. Fields would rapidly be cross-contaminated by pollen flow. However, this describes a hypothetical scenario since Zambia has not allowed growing of GM maize. Open pollinated varieties, which are in use in Chongwe (but in a minority proportion compared to commercial hybrids), have higher outcrossing rates than hybrid cultivars (Sanvido et al. 2008). Farming practices that use an increasing proportion of open pollinated varieties/landraces (termed 'local maize' or 'Gangata' in Chongwe) will thus be more vulnerable to cross-contamination by pollen.

The role of pollen in gene flow is important. However, the vital role of seeds as a vehicle for gene flow must not be forgotten (Dyer et al. 2009). In particular, the human-driven gene flow through intentional and unintentional seed movement is relevant. Farmer's behavior, including management, preservation and selection of seeds have contributed to the diversification of maize landraces (Bellon & Berthaud 2004). Such behaviors, including seed saving and sharing are practices that often go hand in hand with small-scale farming.

The practice of re-using seeds was found to be a common feature among the farmers of Chongwe and was seen as an important part of the local food security and independence. The farmers re-used not only the local maize varieties, but sometimes also commercial hybrid maize varieties. This practice constitutes a link between the informal and the formal seed system. Transgenes would likely be introduced as commercial hybrid seeds in the formal seed system, but might find their way into the informal seed system sooner or later. The combination of pollen flow and the tradition to re-use of seeds would potentially spread and keep transgenes, if introduced, in circulation from year to year. A secondary consequence of such process would be that seeds protected by intellectual property rights (IPR) might be found in local on-farm seed stores, with a risk of patent infringements. The open pollinated local maize varieties would be particularly vulnerable to cross-pollination by transgenes.

The practice of sharing seeds to family members and friends was also common in Chongwe, primarily within the local community. Again, in a GM maize scenario, seed sharing would spread transgenes quickly within the community, and also across communities as exemplified by a farmer sharing seeds up to a distance of 100 km. This would mean that the diffusion of transgenes would not be limited to the community of introduction, but also lead to spread across a larger region. The example illustrates the central role of human agents for gene flow.

Apart from the airborne pollen in anemophilous plants such as maize, it is therefore crucial that decision-makers factor in human driven processes that impact gene flow. Intentionally or unintentionally, humans are the main drivers of gene flow in maize in Mexico (Dyer et al. 2009). The same can be expected in Southern Africa although the scientific data is limited from that region. Small- and subsistence farmers dominate in poor regions and base their agriculture on subsistence practices including seed saving and sharing. These activities must be taken into account in models and understanding of gene flow. Small- and subsistence farmers (commonly referred to among policy makers as the "informal" sector) tend however to be invisible when governmental or commercial analyses and decisions of gene flow are considered. Yet they have to be considered as a substantial component of the biosafety development process.

Conclusion

The agricultural practice under study in the small-scale community of Chongwe is working well in sustaining selection, exchange and re-use of local seed varieties. Thus the farmers uphold the traditional agriculture with its central processes to the local livelihood and food production. It is clear that the studied agricultural practice would not be able to uphold segregation of GM and non-GM maize if transgenes were introduced to that system. Both pollen flow between closely positioned maize fields, and sharing of seeds between farmers represent high rates of gene flow. We need more data to evaluate how representative Chongwe is for other small-scale farming areas and practices, with a focus on field sizes, isolation distances, and proportion of farmers that save/share seeds. For the future, mapping larger areas of local maize field cultivation should enable quantification of transgene maize flow by means of dispersal models for different cultivation patterns and scenarios (see e.g. Reuter et al. 2008).

References

Bannert, M., Stamp, P. (2007) Cross-pollination of maize at long distance. European Journal of Agronomy 27: 44–51.

Beckie, H.J. and Hall, L.M. (2008) Simple to complex: Modelling crop pollen-mediated gene flow. Plant Science 175: 615–628.

Bellon, M.R., Berthaud, J. (2004) Transgenic maize and the evolution of landrace diversity. The importance of farmers' behavior. Plant Physiology 134: 883–888.

Dahl, O.J. (1968) Simula 67 common base language. Norwegian Computing Centre Publication.

Devos, Y., Reheul, D. de Schrijver, A. (2005) The co-existence between transgenic and non-transgenic maize in the European Union: A focus on pollen flow and cross-fertilization. Environmental Biosafety Research 4: 71–87.

Dyer, G.A., Serratos-Hernandez, J.A., Perales, H.R., Gepts, P., Pineyro-Nelson, A., Chavez, A., Salinas-Arreortua, N., Yunez-Naude, A., Taylor, J.E., Alvarez-Buylla, E.R. (2009) Dispersal of Transgenes through Maize Seed Systems in Mexico. PLoS ONE 4: e5743.

Gray, E., Ancev, T., Drynan, R. (2011) Coexistence of GM and non-GM crops with endogenously determined separation. Ecological Economics 70: 2486–2493.

Gustafson, D.I., Brants, I.O., Horak, M.J., Remund, K.M., Rosenbaum, E.W., Soteres, J.K. (2006) Empirical modeling of genetically modified maize grain production practices to achieve European Union labeling thresholds. Crop Science 46: 2133–2140.

Kawashima, S., Nozaki, H., Hamazaki, T., Sakata, S., Hama, T., Matsuo, K., Nagasawa, A. (2011) Environmental effects on long-range outcrossing rates in maize. Agriculture Ecosystems & Environment 142: 410–418.

MAFF (1995) Zambia Seed Production Handbook. Ministry of Agriculture, Food and Fisheries.

Messeguer, J., Penas, G., Ballester, J., Bas, M., Serra, J., Salvia, J., Palaudelmas, M., Mele, E. (2006) Pollen-mediated gene flow in maize in real situations of coexistence. Plant Biotechnology Journal 4: 633–645.

Reuter, H., Böckmann, S., Breckling, B. (2008) Analysing cross-pollination studies in maize. In: Breckling, B., Reuter, H., Verhoeven, R. (eds.): Implications of GM Crop Cultivation at Large Spatial Scales. 47–52.

Sanvido, O., Widmer, F., Winzeler, M., Streit, B., Szerencsits, E., Bigler, F. (2008) Definition and feasibility of isolation distances for transgenic maize cultivation. Transgenic Research 17: 317–335.

Viljoen, C., Chetty, L. (2011) A case study of GM maize gene flow in South Africa. Environmental Science Europe 23: 8–15.

Chapter IV

Causes and effects
Research perspectives and requirements

Teratogenesis by glyphosate based herbicides and other pesticides. Relationship with the retinoic acid pathway

Andrés Carrasco

Lab. Molecular Embryology, School of Medicine UBA, Conicet, Argentina

In South America, the incorporation of genetically modified organisms (GMO) engineered to be resistant to pesticides changed the agricultural model into one dependent on the massive use of agrochemicals (Teubal et al. 2005; Teubal 2009). Different pesticides are used in response to the demands of the global consuming market to control weeds, herbivorous arthropods, and crop diseases.

A recent study using a commercial formulation of glyphosate based herbicides (GBH) showed that treatments with a 1/5000 dilution (430 µM of glyphosate) were sufficient to induce reproducible malformations in embryos of the South African clawed frog Xenopus laevis, a widely used vertebrate model for embryological studies (Paganelli et al. 2010). The phenotypes observed include shortening of the trunk, cephalic reduction, microphthalmy, cyclopia, reduction of the neural crest territory at neurula stages and craniofacial malformations at tadpole stages. In addition GBH inhibits the anterior expression domain of the morphogen Sonic Hedgehog (shh) and reduces the domain of the cephalic marker otx2, prevents the subdivision of the eye field and impairs craniofacial development. Moreover, in recent experiments with another commercial formulation of GBH, the malformations observed before were reproduced in a dose-dependent manner, even at dilutions of 1/500000, which produced developmental abnormalities in 17% of the embryos, without lethality (unpublished results).

It is known that glyphosate penetration through the cell membrane and subsequent intracellular action is greatly facilitated by adjuvants such as surfactants. For this reason, the active principle was also tested by injecting frog embryos with glyphosate alone (between 8 and 12 µM per injected cell). The calculated intracellular concentration for glyphosate injected into embryos was 60 times lower that the glyphosate concentration present in the 1/5000 dilution of the GBH which was used to culture whole embryos. The injection of glyphosate produced similar phenotypes and changes in gene expression, suggesting that the effects are attributable to the active principle of the herbicide.

It is very well known that acute or chronic increase of retinoic acid (RA) levels leads to teratogenic effects during human pregnancy and in experimental models. The characteristic features displayed by RA embryopathy in humans include brain abnormalities such us microcephaly, microphtalmia and impairment of hindbrain development; abnormal external and middle ears (microtia or anotia), mandibular and midfacial underdevelopment, and clefts palate. These craniofacial malformations can be attributed to defects in cranial neural crest cells. An excessive cell death in regions where apoptosis normally takes place may underlie a general mechanism for craniofacial malformations associated to teratogens (Sulik et al. 1988, Clotman et al. 1998).

In fact, an excess of RA signaling is able to down-regulate the expression of shh in the embryonic dorsal midline in Xenopus (Franco et al. 1999, Sharpe & Goldstone 2000). Shh deficiency is associated to the holoprosencephaly syndrome (HPE), a CNS malformation with a frequency of 1/250 of pregnancies and 1/10000 of live births. The HPE is a defect generated by the deficiency of the embryonic dorsal midline, which results in a failure in the division of the brain hemispheres, leading to different grades of craniofacial malformations. Moreover, Shh signaling is also necessary for the development of the cranial neural crest derivatives. In mouse, specific removal of the Shh responsiveness in the neural crest cells that give rise to skeleton and connective tissue in the head, increases apoptosis and decreases proliferation in the branchial arches, leading to facial truncations. In addition Shh signaling from the ventral midline is necessary, as an anti-apoptotic agent, for the survival of the neural epithelium and it is also essential for the rapid and extensive expansion of the early vesicles of the developing midbrain and forebrain (Charrier et al. 2001)

An excess of RA signaling also down-regulates otx2 expression in Xenopus, chicken and mouse embryos (Clotman et al. 1998). Knock-out mice for otx2 lack all the brain structures anterior to rhombomere 3. Interestingly, heterozygous mutants showed craniofacial malformations including loss of the eyes and lower jaw (agnathia). These phenotypes are reminiscent of otocephaly reported in humans and other animals and suggest that otx2 plays an essential role in the development of cranial skeletons of mesencephalic neural crest origin (Matsou et al. 1995, Kimura et al. 1997, Erlich et al. 2000).

All this evidence indicates that RA, otx2 and shh are part of a genetic cascade critical for the development of the brain and craniofacial skeleton of neural crest origin. Glyphosate inhibits the anterior expression of shh, reduces the domain of otx2, prevents the subdivision of the eye field and impairs craniofacial development, resembling aspects of the holoprensecephalic and otocephalic syndromes (Geng & Oliver 2009). Indeed, assays using a RA-dependent gene reporter revealed that GBH treatment increases the endogenous RA activity in Xenopus embryos. Moreover, an antagonist of RA rescued the morphological phenotype produced by GBH. This lead to the conclusion that at least some of the teratogenic effects of GBH were mediated by increased endogenous RA activity in the embryos (Paganelli et al. 2010). This is consistent with the very well known syndrome produced by excess of RA, as described by the epidemiological study of Lammer et al. in humans (Lammer et al. 1985) and in vertebrate embryos (Durston et al. 1989, López & Carrasco 1992, López et al. 1995, Padmanabhan 1998).

In Xenopus embryos, the endogenous activity of retinoids gradually increases during early embryogenesis and is finely regulated in space. Therefore maintaining a normal endogenous distribution of RA is important for axial patterning and organogenesis in vertebrates (Chen et al. 1994, López et al. 1995)

It has been reported that triadimefon, a systemic fungicide with teratogenic effects in rodent models, produces craniofacial malformations in Xenopus laevis by altering endogenous RA signalling (Papis et al. 2007). Arsenic, another endocrine disruptor, also increases RA signaling at low, non-cytotoxic doses, in human embryonic NT2 cells (Davey et al. 2008). In addition, atrazine produces teratogenic effects and decreases the levels of cyp26 transcripts in Xenopus tadpoles, suggesting that this herbicide also disrupts the RA signaling pathway (Lenkowski et al 2008, Lenkowski & McLaughlin 2010). RA signaling is one of the finest pathways to tune up gene regulation during development, and all this evidence raises the possibility that disturbances in RA distribution may be a more general mechanism underlying the teratogenic effects of xenobiotics in vertebrates. Since mechanisms of development are highly conserved in evolution among vertebrates, we would like to stress that they could be useful as very sensitive biosensors to detect undesirable effects of new molecules.

The evidence that links GBH (and potentially other chemicals) to increased activity of the RA signaling pathway might explain the higher incidence of embryonic malformations and spontaneous abortions observed in populations exposed to pesticides.

An important evidence came from the epidemiological study carried out by Benitez-Leite et al. in Paraguay identified 52 cases of malformations in the offspring of women exposed during pregnancy to agrochemicals. The congenital malformations observed include anencephaly, microcephaly, facial defects, myelomeningocele, cleft palate, ear malformations, polydactily, syndactily (Benítez-Leite et al. 2009). These defects are indeed consistent with the well-known and expected syndrome caused by misregulation of the RA pathway.

These conclusions should be taken into account together with the incidence of malformations and cancer in Chaco, an Argentine province with soybean harvest and massive use of glyphosate. Official records reveal a 4-fold increase in developmental malformations in the province and a 3-fold increase of cancer in the locality of La Leonesa in the last decade (Comisión Investigadora 2010).

All this information is extremely worrying because the risk of environmental-induced disruptions in human development is highest during the critical period of gestation (2 to 8 weeks). Moreover, the mature human placenta has been shown to be permeable to glyphosate. After 2.5 hr of perfusion, 15% of administered glyphosate is transferred to the fetal compartment (Poulsen et al. 2009). Indeed, a two-compartment model study suggested that a considerable diffusion of glyphosate into the tissue is reached after intravenous administration in rats. These authors conclude that direct blood concentration is only an average indicator of the presence of the chemical and does not provide evidence about its tissue distribution (Anadón et al. 2009). It is necessary to consider the possibility that very low concentrations (pg/cell and not necessarily evenly distributed to all cells) may be sufficient to cause embryonic lethality (which is consistent with increased frequency of embryonic death and spontaneous abortions) or to modify normal embryonic pattern formation (Antoniou et al. 2011).

References

Anadón, A., Martínez-Larrañaga, M.R., Martínez, M.A., Castellano, V.J., Martínez, M., Martin, M.T. et al. (2009) Toxicokinetics of glyphosate and its metabolite aminomethyl phosphonic acid in rats. Toxicology letters, 190(1): 91–95.

Antoniou, M., Habib, M.E.E.-D.M., Howard, C.V., Jennings, R.C., Leifert, C., Nodari, R.O. et al. (2011) Roundup and birth defects. Is the public being kept in the dark?, Earth Open Source Org. (June 2011). http://www.earthopensource.org/index.php/reports/17-roundup-and-birth-defects-is-the-public-being-kept-in-the-dark

Benítez-Leite, S., Macchi, M.L., Acosta, M. (2009) Malformaciones congénitas asociadas a agrotóxicos, Archivos de Pediatría del Uruguay, 80: 237–247.

Charrier, J.B., Lapointe, F., Le Douarin, N.M., a Teillet, M. (2001) Anti-apoptotic role of Sonic hedgehog protein at the early stages of nervous system organogenesis. Development. 128: 4011–4020.

Chen, Y., Huang, L., Solursh, M. (1994) A concentration gradient of retinoids in the early Xenopus laevis embryo. Developmental Biology 161: 70–76.

Clotman, F., van Maele-Fabry, G., Chu-Wu, L., Picard, J.J (1098) Structural and gene expression abnormalities induced by retinoic acid in the forebrain. Reproductive Toxicology. 12: 169–176.

Comisión Investigadora de contaminantes del agua de la Provincia del Chaco (2010) Informe de la Comisión Investigadora de contaminantes del agua de la Provincia del Chaco, Resistencia, Chaco. Argentina.

Davey, J.C., Nomikos, A.P., Wungjiranirun, M., Sherman, J.R., Ingram, L., Batki, C. et al. (2008) Arsenic as an endocrine disruptor: arsenic disrupts retinoic acid receptor-and thyroid hormone receptor-mediated gene regulation and thyroid hormone-mediated amphibian tail metamorphosis. Environmental Health Perspectives. 116: 165–172.

Durston, A.J., Timmermans, J.P., Hage, W.J., Hendriks, H.F., de Vries, N.J., Heideveld, M. et al. (1989) Retinoic acid causes an anteroposterior transformation in the developing central nervous system. Nature. 340: 140–144.

Erlich, M.S., Cunningham, M.L., Hudgins, L. (2000) Transmission of the dysgnathia complex from mother to daughter. American Journal of Medical Genetics. 95: 269–274.

Franco, P.G., Paganelli, A.R., López, S.L., Carrasco, A.E. (1999) Functional association of retinoic acid and hedgehog signaling in Xenopus primary neurogenesis. Development. 126: 4257–4265.

Geng, X., Oliver, G. (2009) Pathogenesis of holoprosencephaly, J. Clin. Invest. 119 1403-1413.

Kimura, C., Takeda, N., Suzuki, M., Oshimura, M., Aizawa, S., Matsuo, I. (1997) Cis-acting elements conserved between mouse and pufferfish Otx2 genes govern the expression in mesencephalic neural crest cells. Development. 124: 3929–3941.

Lammer, E.J., Chen, D.T., Hoar, R.M., Agnish, N.D., Benke, P.J., Braun, J.T. et al. (1985) Retinoic acid embryopathy. New England Journal of Medicine 313: 837–841.

Lenkowski, J.R., McLaughlin, K.A. (2010) Acute atrazine exposure disrupts matrix metalloproteinases and retinoid signaling during organ morphogenesis in Xenopus laevis, Journal of Applied Toxicology. 30: 582–589.

Lenkowski, J.R., Reed, J.M., Deininger, L., McLaughlin, K.A. (2008) Perturbation of organogenesis by the herbicide atrazine in the amphibian Xenopus laevis. Environmental Health Perspectives. 116: 223–230.

López, S.L., Carrasco, A.E. (1992) Retinoic acid induces changes in the localization of homeobox proteins in the antero-posterior axis of Xenopus laevis embryos. Mechanisms of Development. 36: 153–164.

López, S.L., Dono, R. Zeller, R., Carrasco, A.E. (1995) Differential effects of retinoic acid and a retinoid antagonist on the spatial distribution of the homeoprotein Hoxb-7 in vertebrate embryos. Developmental Dynamics. 204: 457–471.

Matsuo, I., Kuratani, S., Kimura, C., Takeda, N., Aizawa, S. (1995) Mouse Otx2 functions in the formation and patterning of rostral head. Genes & development. 9: 2646–2658.

Padmanabhan, R. (1998) Retinoic acid-induced caudal regression syndrome in the mouse fetus. Reproductive Toxicology 12: 139–151.

Paganelli, A., Gnazzo, V., Acosta, H., López, S.L., Carrasco, A.E. (2010) Glyphosate-Based Herbicides Produce Teratogenic Effects on Vertebrates by Impairing Retinoic Acid Signaling. Chemical Research in Toxicology 23: 1586–1595.

Papis, E., Bernardini, G., Gornati, R., Menegola, E., Prati, M. (2007) Gene expression in Xenopus laevis embryos after Triadimefon exposure. Gene Expression Patterns. 7: 137142.

Poulsen, M.S., Rytting, E., Mose, T., Knudsen, L.E. (2009) Modeling placental transport: correlation of in vitro BeWo cell permeability and ex vivo human placental perfusion. Toxicology in Vitro. 23: 1380–1386.

Sharpe, C., Goldstone, K., (2000) Retinoid signalling acts during the gastrula stages to promote primary neurogenesis. International Journal of Developmental Biology 44: 463–470.

Sulik, K.K., Cook, C.S., Webster, W.S. (1988) Teratogens and craniofacial malformations: relationships to cell death. Development. 103 Suppl: 213–231.

Teubal, M. (2009) Expansión del modelo sojero en la Argentina. De la producción de alimentos a los commodities. In: Lizarraga, P., Vacaflores, C. (Eds.) La persistencia del campesinado en América Latina, Comunidad de Estudios JAINA, Tarija: pp. 161–197.

Teubal, M., Domínguez, D., Sabatino, P. (2005) Transformaciones agrarias en la Argentina. Agricultura industrial y sistema agroalimentario. In: Giarracca, N., Teubal, M. (Eds.) El campo argentino en la encrucijada. Estrategias y resistencias sociales, Ecos en la ciudad, Alianza Editorial, Buenos Aires: pp. 37–78.

Human cell toxicity of pesticides associated to wide scale agricultural GMOs

Robin Mesnage, Steeve Gress, Nicolas Defarge & Gilles-Eric Séralini

University of Caen, Institute of Biology, CRIIGEN and Risk Pole, MRSH-CNRS, Caen Cedex, France

Agricultural genetically modified (GM) plants are essentially plants which contain pesticides, because they were designed to tolerate or produce pesticides. In 2011, GM crops reached 160 million hectares, with 59 % of herbicide tolerance (mainly Roundup) mostly in soybean, maize, canola, cotton, 15 % of insecticide producing varieties and 26 % combining both traits (James 2011). We characterized cellular side effects of these pesticide residues on non-target human cells. We summarized here our findings:

Glyphosate-based herbicides toxicity

Roundup (R) was highly toxic on human cells, from 10-20 ppm, far below agricultural dilutions. This occurred on hepatic (HepG2, Hep3B and embryonic (HEK293) as well on placental (JEG3) cell lines, but also on human placental extracts, primary umbilical cord cells (HUVEC) and freshly isolated testicular cells (Richard et al. 2005; Benachour et al. 2007; Benachour & Seralini 2009; Gasnier et al. 2010; Clair et al. 2012). All formulations cause total cell death within 24 h, through an inhibition of the mitochondrial succinate dehydrogenase activity, and necrosis, through the release of cytosolic adenylate kinase measuring membrane damage. They also induced apoptosis through the activation of enzymatic caspases 3/7 activities. Most importantly, the R commercialized formulation is always more toxic than the active principle alone, the glyphosate (G). These effects were more dependent on the formulation and thus adjuvants content than on the G concentration. We recently measured compositions and effects of 9 G-based formulations and identified ethoxylated adjuvants (commonly called POEA) as the active principle of cytotoxicity (Messnage et al. 2012a). However, these are considered as inert diluents in international regulations and are not taken into account for chronic effects which are insufficiently tested, and only with G in pre-commercial testing. We previously underlined this loophole (Mesnage 2010). Long term feeding and reproductive trials with pesticides are the only tests long enough to reveal a potential endocrine disruption which was consequently never studied for R until recently (Seralini et al. 2012), however it was for G by itself.

We investigated it by measuring androgen to estrogen conversion by aromatase activity and mRNA on placental human cells and showed that G interacts with the active site of the purified enzyme (Richard et al. 2005). Both parameters were disrupted at sub-agricultural doses within 24 h. We also observed a human cell endocrine disruption from 0.5 ppm on the androgen receptor in transfected cells, and then from 2 ppm the transcriptional activities on both estrogen receptors which were also inhibited (Gasnier et al. 2009). Aromatase transcription and activity were disrupted from 10 ppm on HepG2. On freshly isolated rat testicular cells, low non-toxic concentrations of R and G (1 ppm) induced a testosterone decrease by 35 % (Clair et al. 2012). This is expected to occur in human cells which are fitted with the same steroidogenic equipment.

G-based formulations are claimed to have been extensively studied by industry and regulatory agencies and are considered as one of the safest pesticides (Williams et al. 2000). This allowed the establishment of high maximum residue limits (MRL) for GM food/feed (up to 400 ppm). For instance, 20 ppm of G are authorized in GM soy and this MRL is in the range of concentrations typically found in a GM soy harvest. In the light of our results, the safety of these thresholds is clearly challenged.

Insecticidal toxins (Bt) toxicity

Modified toxins from *Bacillus thuringiensis* are Cry proteins forming pores in insect cell membranes (Then 2010). They are claimed and believed to be inert on non-target species. We have tested for the very first time Cry1Ab and Cry1Ac modified Bt toxins (10 ppb to 100 ppm) on the HEK293 cell line, as well as their combined actions with R, within 24 h, on three biomarkers of cell death: measurements of mitochondrial succinate dehydrogenase, adenylate kinase release by membrane alterations and caspases 3/7 inductions (Mesnage et al. 2012b). Modified Cry1Ab caused cell death from 100 ppm. For Cry1Ac, under such conditions, no effects were detected. In vivo implications should be now assessed, as Cry1Ab does not appear to be proved as an insect specific toxin.

Combined toxicity

In the new growing generation with stacked traits, G-based herbicides (like R) residues are present in the R-tolerant edible plants and mixed with modified Bt insecticidal toxins that are produced by the GM plants themselves. However, the toxicology of mixtures cannot be fully understood without knowing the combined toxicity of the various compounds of the formulations. In some in vitro conditions, G and its adjuvant synergistically damage cell membranes in a similar manner to R (Benachour & Seralini 2009). R adjuvants change human cell permeability and amplify toxicity induced already by G, through apoptosis and necrosis. The real threshold of G toxicity must take into account the presence of adjuvants but also G metabolism and time-amplified effects or bioaccumulation. For the mixtures of Bt toxins and R, the only measured significant combined effect was that modified Cry1Ab and Cry1Ac reduced caspases 3/7 activations induced by R; this could delay the activation of apoptosis and impact on necrosis.

There was the same tendency for adenylate kinase activity and succinate dehydrogenase activity measures. Pesticides have to be tested together, 26 % of agricultural GMOs are indeed stacked events.

We also reviewed 19 studies of mammals fed with commercialized GMOs (Seralini et al. 2011). Meta-analysis of all biochemical disruptions indicated liver and kidney problems as end points of GMO diet effects. These are the major reactive organs in case of food chronic intoxication, and several contingent factors suggested that pesticide residues may be involved in the pathological features. All together, our results raise new questions in the risk assessment of food and feed derived from genetically engineered plants.

References

Benachour, N., Seralini, G.E. (2009) Glyphosate formulations induce apoptosis and necrosis in human umbilical, embryonic, and placental cells. Chem Res Toxicol 22: 97–105.

Benachour, N., Sipahutar, H., Moslemi, S., Gasnier, C., Travert, C., Seralini, G.E. (2007) Time- and dose-dependent effects of roundup on human embryonic and placental cells. Arch Environ Contam Toxicol 53: 126–133.

Clair, E., Mesnage, R., Travert, C., Seralini, G.E. (2012) A glyphosate-based herbicide induces necrosis and apoptosis in mature rat testicular cells in vitro, and testosterone decrease at lower levels. Toxicol In Vitro 26: 269–279.

Gasnier, C., Benachour, N., Clair, E., Travert, C., Langlois, F., Laurant, C., Decroix-Laporte, C., Seralini, G.E. (2010). Dig1 protects against cell death provoked by glyphosate-based herbicides in human liver cell lines. J Occup Med Toxicol 5: 29.

Gasnier, C., Dumont, C., Benachour, N., Clair, E., Chagnon, M.C., Seralini, G.E. (2009) Glyphosate-based herbicides are toxic and endocrine disruptors in human cell lines. Toxicology 262: 184–191.

James, C. (2011) Global Status of Commercialized Biotech/GM Crops: 2011. ISAAA Brief 43.

Mesnage, R., Clair, E., Séralini, G.E. (2010) Roundup in genetically modified plants: Regulation and toxicity in mammals. Theorie in der Ökologie 16: 31–33.

Mesnage, R., Bernay, B., Séralini, G.E. (2012a) Ethoxylated adjuvants of glyphosate-based herbicides are active principles of human cell toxicity. Toxicology. doi:10.1016/j.tox.2012.09.006

Mesnage, R., Clair, E., Gress, S., Then, C., Szekacs, A., Seralini, G.E. (2012b) Cytotoxicity on human cells of Cry1Ab and Cry1Ac Bt insecticidal toxins alone or with a glyphosate-based herbicide. J Appl Toxicol. doi: 10.1016/j.tox.2012.09.006

Richard, S., Moslemi, S., Sipahutar, H., Benachour, N., Seralini, G.E. (2005). Differential effects of glyphosate and roundup on human placental cells and aromatase. Environ Health Perspect 113: 716–720.

Seralini, G.E., Mesnage, R., Clair, E., Gress, S., de Vendomois, J., Cellier, D. (2011) Genetically modified crops safety assessments: present limits and possible improvements. Environmental Sciences Europe 23: 10.

Séralini, G.E., Clair, E., Mesnage, R., Gress, S., Defarge, N., Malatesta, M., Hennequin, D., Spiroux de Vendômois, J. (2012) Long term toxicity of a Roundup herbicide and a Roundup-tolerant genetically modified maize. Food Chem Toxicol. 50: 4221–4231.

Effect of extreme climatic conditions on Bt toxin concentration in transgenic maize

Miluse Trtikova[1], Matthias S. Meier[2] & Angelika Hilbeck[1]

[1]ETH Zurich, Institute of Integrative Biology; [2]FiBL, Research Institute of Organic Agriculture, Frick; Switzerland

Extended Abstract[1]

Numerous studies have shown that transgenic Bt maize plants do not produce Bt toxin at a uniform level (Abel & Adamczyk 2004; Nguyen & Jehle 2007; Szekacs et al. 2010). It was suggested that the toxin expression might be affected by the plant genetic background and various environmental factors, including climate. According to IPCC report (2012), the frequency of extreme climatic events, such as droughts or heavy rains, is likely to increase in the future. However, it is not known whether such climatic events influence the Bt toxin expression in transgenic maize. Therefore, we measured Bt toxin concentration in the leaves of Bt maize exposed to drought in combination with high temperatures and waterlogging in combination with low temperatures.

First, we grew two MON810 varieties (PAN6Q-321B and PAN6Q-308B) for five to six weeks under optimal conditions (20–25 °C, watered regularly). Then, we exposed the plants for two weeks to two different treatments, simulating either hot/dry (21–30 °C, watered only sparsely) or cold/wet (13–16 °C, waterlogged) conditions. The upper leaves were sampled before and after the plant exposure to extreme climatic conditions and the Bt toxin content was quantified using ELISA.

Although the data are still being evaluated, the preliminary results indicate that waterlogging in combination with low temperatures may lead to a decrease in Bt toxin concentration in the leaves of transgenic Bt maize. This could have direct consequences for both target and non-target insects, especially in terms of insect resistance development and environmental risk assessment. If the plants produce lower amount of Bt toxin, they might pose less risk to the non-target insects, but at the same time the survival of the target insects could also increase and they might develop resistance to Bt toxin.

---

1 A full paper is in preparation.

References

Abel, C.A., Adamczyk, J.J. (2004) Relative concentration of Cry1A in maize leaves and cotton bolls with diverse chlorophyll content and corresponding larval development of fall armyworm (Lepidoptera: Noctuidae) and southwestern corn borer (Lepidoptera: Crambidae) on maize whorl leaf profiles. Journal of Economic Entomology 97: 1737–1744.

Field, C.B., Barros, V., Stocker, T.F., Qin, D., Dokken, D.J., Ebi, K.L., Mastrandrea, M.D., Mach, K.J., Plattner, G.-K., Allen, S.K., Tignor, M., Midgley, P.M., (eds) (2012) IPCC, 2012: Summary for Policymakers. In: Managing the Risks of Extreme Events and Disasters to Advance Climate Change Adaptation. A Special Report of Working Groups I and II of the Intergovernmental Panel on Climate Change. Cambridge University Press, Cambridge, UK, and New York, USA. 1–19.

Nguyen, H.T., Jehle, J.A. (2007) Quantitative analysis of the seasonal and tissue-specific expression of Cry1Ab in transgenic maize Mon810. Journal of Plant Diseases and Protection 114: 82–87.

Szekacs, A., Lauber, E., Takacs, E., Darvas, B. (2010) Detection of Cry1Ab toxin in the leaves of MON 810 transgenic maize. Analytical and Bioanalytical Chemistry 396: 2203–2211.

Cry1Ab toxin quantification in MON 810 maize

András Székács

Central Food Science Research Institute, Budapest, Hungary

Extended Abstract[1]

Quantification of the expressed transgenic Cry1Ab toxin in MON 810 maize reveals information on several crucial aspects regarding the agrotechnological application of this genetic event. Although resulting in the same trypsine-cleaved toxin form (activated Cry1Ab toxin, 65 kDa) in the insect midgut, the plant-expressed toxin protein (preactivated Cry1Ab toxin, 91 kDa) is a truncated version of the corresponding bacterial toxin (Cry1Ab protoxin, 131 kDa). This difference has both analytical and regulatory consequences.

As commercial enzyme-linked immunosorbent assays (ELISAs), widely used for toxin quantification, are devised against the bacterial protoxin, and use the bacterial protoxin as analytical standard, they reportedly underestimate Cry1Ab toxin content in MON 810 maize (Székács et al. 2010a). As a result, data on Cry1Ab content that are based on such experimental conditions (including the documentation by the variety owner, Monsanto) are subject to correction by the factor of the preactivated toxin / protoxin cross-reactivity value. Moreover, the Cry1Ab preactivated toxin produced by MON 810 maize is not a registered *Bacillus thuringiensis* (Bt) based bioinsecticide ingredient in the formal sense; its authorization would require complete toxicological evaluation. Moreover, commercial availability of the plant expressed preactivated Cry1Ab toxin as analytical standard and of commercial ELISA systems directed against the preactivated toxin is urged (Székács & Darvas 2012b).

Analytical detection of Cry1Ab is hindered by substantial plant-to-plant and within plant tissue variation in Cry1Ab content (Székács et al. 2010b), as well as by potential variability among results obtained in different laboratories (Székács et al. 2012). Yet, data on Cry1Ab content unequivocally indicate the MON 810 maize (as Bt crops in general) release orders of magnitude more bioavailable toxin into the environment than the amount released with a treatment with a Bt based bioinsecticide (e.g., DIPEL) (Székács et al. 2005; Székács & Darvas 2012a). As for their environmental fate, in contrast to the rapid breakdown of the bacterial Cry1Ab protoxin, the plant-expressed preactivated Cry1Ab toxin protein shows persistence in the stubble, being protected from decomposition within the plant cells.

---

1 A full paper is in preparation.

And finally, although MON 810 maize has been advocated to be included in integrated pest management (IPM) practices, it cannot fulfil the main ecological principle of IPM, e.g. that protection measures should be timed only to the period when pest damage exceeds the critical level. Therefore, in spite of the environmentally mild characteristics of its active ingredient is, MON 810 maize does not comply with IPM (Székács & Darvas 2012).

References

Székács, A., Darvas, B. (2012a) Environmental and ecological aspects of first generation genetically modified crops regarding their impacts in a European maize producer country. International Journal of Environmental Protection. 2 (5): 9–15.

Székács, A., Darvas B. (2012b) Comparative aspects of Cry toxin usage in insect control. In: Ishaaya, I., Palli, S.R., Horowitz, R. (eds.) Advanced technologies for managing insect pests, Springer-Verlag, Berlin, in press. (DOI 10.1007/978-94-007-4497-4_10)

Székács, A., Juracsek, J., Polgár, L. A., Darvas, B. (2005) Levels of expressed Cry1Ab toxin in genetically modified corn DK-440-BTY (YIELDGARD) and stubble. FEBS J., 272 (Suppl.1.): 508.

Székács, A., Lauber, É., Juracsek, J., Darvas, B. (2010) Cry1Ab toxin production of MON 810 transgenic maize. Environmental Toxicology and Chemistry. 29 (1): 182–190.

Székács, A., Lauber, É., Takács, E., Darvas, B. (2010) Detection of Cry1Ab toxin in the leaves of MON 810 transgenic maize. Analytical & Bioanalytical Chemistry. 396 (6): 2203–2211.

Székács, A., Weiss, G., Quist, D., Takács, E., Darvas, B., Meier, M., Swain, T., Hilbeck, A. (2012) Inter-laboratory comparison of Cry1Ab toxin quantification in MON 810 maize by enzyme-immunoassay. Food and Agricultural Immunology. 23 (2): 99–121.

Are frogs and toads affected by complementary herbicides of GM crops?

Norman Wagner[1], Wolfram Reichenbecher[2], Hanka Teichmann[2], Beatrix Tappeser[2] & Stefan Lötters[1]

[1] Department of Biogeography, Trier University; [2] Federal Agency for Nature Conservation, Bonn, Germany

Abstract

Since the introduction of genetically modified herbicide-resistant crops in 1996, the use of complementary herbicides, especially glyphosate-based herbicides (GBH), continues to grow at the global scale. By reviewing relevant literature, we investigated to what extent anuran amphibians are affected by GBH. The impact of GBH on these vertebrates depends on the herbicide formulation, the species considered and its life-stage. Only little is known about the glyphosate (GLY) concentrations prevailing in the animals' habitats and virtually nothing is known about environmental pollution by further substances contained in herbicide formulations. Therefore, GLY levels deduced from limited measurements can only be seen as approximations for contamination of anuran habitats with GBH. Co-stressors mostly increase negative effects of GBH to anurans and can render apparently harmless GBH concentrations harmful. We recommend to create an anuran test battery which should be mandatory with each pesticide formulation for registration. Amphibian populations are declining due to multiple stressors, but especially due to over-application and missing buffer strips GBH use may become a main stressor for populations in intensively used agrarian landscapes. We strongly recommend standardized monitoring of both amphibian populations and GBH in the environment.

Introduction

Glyphosate (GLY) is the active ingredient in many broad-spectrum herbicide formulations. Environmental risk assessment of GLY and glyphosate-based herbicides (GBH) is highly relevant as the application of GBH has drastically increased at the global scale. This is mainly due to the success of genetically modified crops with herbicide resistance (Duke & Powles 2008). Based on results from specific amphibian studies, several researchers claim amphibians, especially anuran larvae, to be particularly sensitive to GLY and GBH (e.g. Relyea 2011) and, compared to other animals, terrestrial life-stages of amphibians seem to be at higher risk due to dermal uptake. However, amphibians do not belong to the standard test organisms for pesticide approval. Results from (larval) teleost fish and aquatic invertebrates are usually transferred to aquatic amphibian larvae and those from mammals and birds to terrestrial amphibian adults (Aldrich 2009; Relyea 2011).

Pesticide ingredients are divided into 'active' and 'inert' ingredients (Cox & Surgan 2006). Standard tests revealed that the active ingredient (GLY) is usually less toxic than the commercial formulation (GBH), mainly due to the surfactants, for instance the most common surfactant POEA (polyethoxylated tallow amine). Because of inconsistency in conclusions about the effects of GLY and GBH on amphibians in the environment, we reviewed relevant literature from the past 13 years. The following results and conclusions mainly refer to anuran amphibians (frogs, toads), especially to their larvae (tadpoles).

Toxicological effects

During their life-cycle, most anurans have an aquatic (egg, embryo and tadpole) and a terrestrial phase (metamorph, juvenile and adult). We therefore separated aquatic from terrestrial life-stages among our findings.

Aquatic life-stages – acute toxicity
As for aquatic standard test organisms, GBH are usually more toxic to tadpoles than GLY alone. This is due to the surfactants (Mann & Bidwell 1999). GBH with POEA and other tallowamine surfactants are 'highly toxic' to tadpoles of some species and so are GBH with other surfactants (Relyea 2011). However, it should be noted that most GBH are 'moderately toxic' and some 'slightly toxic' to even 'practically non-toxic' to most tadpoles (Mann & Bidwell 1999; Howe et al. 2004). Consequently, acute toxicity of GBH exposure to tadpoles is formulation- (i.e. surfactant) and species-dependent.

Teratogenicity
The few studies which investigated effects of GLY and GBH on embryonic development all used different methodological designs making it difficult to compare. Therefore, consistent conclusions about teratogenic effects are still missing.

Aquatic life-stages – chronic effects
Sublethal concentrations of some GBH bear the potential to disrupt larval development. Also natural stressors such as competition or dry-out of ponds can lead to differences in body mass at metamorphosis and time to metamorphosis. An evaluation of long-term effects, however, is difficult since specific studies on natural amphibian populations or life-cycle tests have not been conducted.

Based on the results of some studies, the USEPA (2012) calculated ecotoxicological endpoints (NOEC, LOEC) for chronic effects of GLY and GBH on amphibians. A comparison of these studies is hardly possible because they differ in design and the sublethal effects investigated.

Terrestrial life-stages
None of the studies addressing acute effects referred to chronic effects. Direct over-spray of individuals with some GBH can pose direct health effects while other GBH apparently do not cause any acute effect with recommended application rates. Furthermore, effects are species-dependent. Standard procedures for direct over-spray of individuals or just their skin (Quaranta et al. 2009) enable assessing the risk of dermal uptake during terrestrial life-stages.

Interactions of GLY and GBH with other stressors

Existing studies only refer to tadpoles. Interactions of GLY, GBH and co-stressors present in larval amphibian habitats have been studied under laboratory conditions and with the help of mesocosms (simulating natural aquatic communities). The presence of abiotic and biotic co-stressors such as other pesticides, predators, competition, higher pH and UV-B radiation in addition to GLY and GBH may enhance their adverse effects. With pathogenic organisms some converse effects have been observed when combined with GBH. These may be related to effects of GBH on the pathogenic organisms. However, the negative effects on the tadpoles prevail.

Concentrations of GLY and GBH in amphibian habitats

Maximum concentrations are relevant because acute toxic effects are often observed within few hours after application. In general, information on GLY concentrations in the environment is sparse. One reason is that GLY analysis is expensive. GLY concentrations in aquatic amphibian habitats (mainly small ponds) are widely lacking. Data from other surface waters (mainly rivers, lakes) should be considered as minimum contamination. Data from terrestrial habitats cannot be considered due to lack of specific amphibian studies on uptake via contaminated soil or plant material.

Maximum GLY concentrations in surface waters can be divided into three different groups: concentrations that were
- measured directly in the environment with unknown degradation state so that higher peak concentrations are likely (up to 0.7 mg acid equivalent (a.e)/L next to GM soybean cultivations after heavy rainfall; Peruzzo et al. 2008);
- measured in field studies directly after experimental herbicide application (up to 3.1 mg a.e./L after aquatic use; Trumbo 2005) and up to 1.95 mg a.e./L after aerial application at recommended application rates (Thompson et al. 2004);
- estimates of worst-case scenarios (up to 7.6 mg a.e./L; Mann & Bidwell 1999). A direct comparison of concentrations from the three groups is not possible and GLY levels can only be regarded as approximations for contamination with GBH because there is no information about further substances in the herbicide formulations.

Effects at the population level

Pesticide use is one among several reasons supposed to cause on-going global amphibian declines and extinctions (Boone et al. 2007). Its specific role remains difficult to assess because field data remain sparse. Abnormal population changes have been suggested to result from multiple interacting causes. No precise information is available with regard to the past, current or future contribution of GBH to amphibian decline. The increasing use of GBH cannot be made responsible for the gross of amphibian declines during the last 50 years, for following reasons. The majority of declines in Western Europe occurred in the 1960s to 1970s (Houlahan et al. 2000), probably due to intensification in agriculture but also general habitat destruction and contamination by the growth of industry. GLY marketing has only started in 1974 (Dill et al. 2010). Dramatic declines and extinctions of amphibians have been witnessed over the last three decades, but mainly in pristine and remote areas of the tropics. An important role of GBH use can be ruled out here, but other potential causes have been identified like habitat destruction and emerging infectious diseases such as the amphibian chytrid fungus (Mendelson et al. 2006).

Scientific studies and (limited) information on environmental concentrations indicate that worst-case conditions, missing buffer strips and especially (illegal) over-application of certain GBH may lead to local population declines, especially of previously affected populations in intensively used agrarian landscapes. In Argentina large areas (up to 89%) covering the habitats of anuran species (partly endemic) are covered with HR soybean cultivation (Lajmanovich et al. 2010).

Hence, more research is required to answer whether and to what extent GBH use affects amphibian populations.

Conclusions and future research issues

Safety factors and standard test procedures in pesticide regulation largely account for the acute effects on tadpoles, but they consider neither teratogenic or chronic effects nor dermal uptake by amphibian adults and juveniles.

In the context of GBH regulation the risks for amphibians and also other organisms are assessed prospectively. Real risks to amphibians (tadpoles) are best assessed at local level with the specific GBH known.

Three main issues should be considered for future research. We recommend to (i) perform at least one amphibian- (anuran-) specific and mandatory test battery for pesticide registration, (ii) conduct a standardized long-term monitoring of GBH, not only GLY, in amphibian habitats and (iii) create a standardized long-term monitoring of natural amphibian populations aiming at the detection of abnormal local population changes.

References

Aldrich, A. (2009) Empfindlichkeit von Amphibien gegenüber Pflanzenschutzmitteln. Agrarforschung 16: 466–471.

Boone, M.D., Cowman, D., Davidson, C., Hayes, T., Hopkins, W., Relyea, R.A., Schiesari L., Semlitsch, R. (2007) Evaluating the Role of Environmental Contamination in Amphibian Population Declines. In: Gascon, C., Collins, J.P., Moore, R.D., Church, D.R., McKay, J.E., Mendelson, J.R. III (eds.) IUCN Amphibian Conservation Action Plan. Gland & Cambridge, IUCN/SSC Amphibian Specialist Group. 32–36.

Cox, C., Surgan, M. (2006) Unidentified inert ingredients in pesticides: Implications for human and environmental health. Environmental Health Perspectives 114: 1803–1806.

Dill, G.M., Sammons, R.D., Feng, P.C.C., Kohn, F., Kretzmer, K., Mehrsheikh, A., Bleeke, M., Honegger, J.L., Farmer, D., Wright, D., Haupfear, E.A. (2010) Glyphosate: Discovery, Development, Applications, and Properties. In: Mandula, N. (ed.) Glyphosate Resistance in Crops and Weeds. Hoboken, John Wiley & Sons. 1–33.

Duke, S.O., Powles, S.B. (2008) Glyphosate: a once-in-a-century herbicide. Pest Management Science 64: 319–325.

Houlahan, J.E., Findlay, C.S., Schmidt, B.R., Meyer, A.H., Kuzmin, S.L. (2000) Quantitative evidence for global amphibian population declines. Nature 404: 752–755.

Howe, C.M., Berrill, M., Pauli, B.D., Helbing, C.C., Werry, K., Veldhoen, N. (2004) Toxicity of glyphosate-based pesticides to four North American frog species. Environmental Toxicology and Chemistry 23: 1928–1938.

Lajmanovich, R.C., Peltzer, P.M., Junges, C.M., Attademo, A.M., Sanchez, L.C., Bassó, A. (2010) Activity levels of B-esterases in the tadpoles of 11 species of frogs in the middle Paraná River floodplain: Implication for ecological risk assessment of soybean crops. Ecotoxicology and Environmental Safety 73: 1517–1524.

Mann, R.M., Bidwell, J.R. (1999) The toxicity of glyphosate and several glyphosate formulations to four species of southwestern Australian frogs. Archives of Environmental Contamination and Toxicology 36: 193–199.

Mendelson, J.R., Lips,K.R., Gagliardo, R.W., Rabb, G.B., Collins, J.P., Diffendorfer, J.E., Daszak, P., Ibanez, R., Zippel, K.C., Lawson, D.P., Wright, K.M., Stuart, S.N., Gascon, C., da Silva, H.R., Burrowes, P.A., Joglar, R.L., La Marca, E., Lötters, S., et al. (2006) Biodiversity - Confronting amphibian declines and extinctions. Science 313: 48–48.

Peruzzo, P.J., Porta, A.A., Ronco, A.E. (2008) Levels of glyphosate in surface waters, sediments and soils associated with direct sowing soybean cultivation in north pampasic region of Argentina. Environmental Pollution 156: 61–66.

Quaranta, A., Bellantuono, V., Cassano, G., Lippe, C. (2009) Why amphibians are more sensitive than mammals to xenobiotics. PLoS ONE 4: e7699.

Relyea, R.A. (2011) Amphibians Are Not Ready for Roundup®. In: Elliott, J.E., Bishop, C.A., Morrissey, C.A. (eds.) Wildlife Ecotoxicology Vol 3 – Emerging Topics in Ecotoxicology. New York, Springer. 267–300.

Thompson, D.G., Wojtaszek, B.F., Staznik, B., Chartrand, D.T., Stephenson, G.R. (2004) Chemical and biomonitoring to assess potential acute effects of Vision® herbicide on native amphibian larvae in forest wetlands. Environmental Toxicology and Chemistry 23: 843–849.

Trumbo, J. (2005) An assessment of the hazard of a mixture of the herbicide Rodeo® and the non-ionic surfactant R-11® to aquatic invertebrates and larval amphibians. California Fish and Game 91: 38–46.

USEPA (United States Environmental Protection Agency) (2012) http://cfpub.epa.gov/ecotox

Transgenic evolution and ecology are proceeding

Broder Breckling[1,2]

[1] Landscape Ecology, University of Vechta; [2] University of Bremen, UFT Centre of Environmental Research and Sustainable Technology, Bremen; Germany

Introduction

The agronomic use of genetic engineering introduces plants with properties that do not occur naturally or through conventional breeding. The new traits potentially end up in the gene pool of the species and subsequently become part of the population genetic dynamics.

Unlike the results of conventional breeding, genetically engineered plants can be patent protected. They are accordingly subjected to intellectual property rights. The patent holder has rights and responsibilities concerning the role that his property plays in cultivation and beyond. Like any other plant species in cultivation, genetically engineered plants do not only occur in the intended places but interact with the surrounding natural environment (Ellstrand 2003). This has to be considered in risk assessment before the release of GM plants. According to the biology of the particular species, genetic exchange with related, feral species or escape from cultivation is possible and has been predicted (Marvier & van Acker 2005).

Since in some countries genetically modified plants are cultivated to a relevant extent in agriculture, it is worth while to ask and to assess to what extent unintended occurrences of GM plants in ecological contexts actually emerged and whether anticipated risks become manifest. Systematic and representative investigations of unintended occurrences of GMO and their long-term implications are still lacking.

It is important to keep track of ecological developments which emerge self-organised and involve transgenes. It is only possible to do this on a scientific level if specific funding is available (Ober 2009; Graef et al. 2012). Otherwise, ecological dynamics with potentially considerable sustainability implications remain undocumented. An interesting example was the Triffid case (Schmidt & Breckling 2010): For nearly ten years it had escaped the attention of scientific documentation as well as regulatory oversight that a genetically engineered flax variety which had been on the market practically only in one growing season in Canada had caused impurities. This finally led to severe export problems on national and international markets when it was discovered a bout a decade later.

Transgenic traits – not on leash

From the beginning of GM crop cultivation it was obvious that transgenes would not remain restricted to agricultural areas and enter natural ecosystems (Marvier & van Acker 2005).

When the particular GM plants reproduce outside agriculturally managed areas, it is apparently inevitable that they subsequently undergo evolutionary dynamics. After almost two decades of growing transgenic crops in environmentally open systems, investigations of evolutionary processes basing on transgenes become more and more important. This topic has received by far too little attention in international research and requires representative ecological investigations. It is a current research deficit to understand how social and agronomic drivers interact with ecological dynamics of transgenic crops on larger scales.

Evolutionary implications

Evolutionary processes can amplify rare events to macroscopic scales.
Evolutionary dynamics connect large numbers on the population level with singularities on the molecular scale (mutations, recombinations, crossing over, etc.). Events with very low probability have a chance to occur because of the large number of repetitions which are usually involved in population processes. Depending on the particular environmental conditions (in case of a systemic fit) the reproduction of organisms enables self-amplification across several orders of magnitude and large scale dispersal. Genetic drift can cause the fixation of genes on a pure random basis particularly in small populations which later may expand.

In this context it is also noteworthy that the fitness of new genomic constituents can not be calculated in absolute terms. Fitness depends not only on the phenotype but also how it conforms to the environment and its future changes. Predictions whether particular transgenes could become the starting point for unintended developments are notoriously difficult. Prospective fitness anticipations in the given context are practically impossible. Since the potential future development of transgenes in native populations is not predictable to a relevant extent, the introduction of transgenes into natural populations must be considered as undesirable. The following examples illustrate that crop specific and environmental considerations are required.

Example 1: Mexican Maize adaptation

The sustainability of traditional maize cultivation in Mexico involves ongoing adaptation dynamics – currently overlaid by transgenes.
Traditional cultivation of maize landraces in Mexico (Acevedo et al. 2011) gives room for self-organised genetic exchange and environmental adaptation of traditional varie-

ties. Due to GM cultivation in the neighbouring USA, from which grains are imported to Mexico enabling uncontrolled sowing of imported grains, the traditional system is exposed to an unknown extent to GM impurities. The rate of gene flow between traditionally managed fields is relatively high because the landraces have a high rate of outcrossing. The high gene flow rate in traditional („open pollinated") non-hybrid varieties synchronously enabled an adaptation to the environment, to pests, and makes farmers selection efficient: This can be understood as an evolved condition of resilience in the local social ecological system.

Transgenes in Mexican maize landraces have been first discovered in 2001 (Chapela & Quist). Their presence was independently confirmed (Pineyro-Nelson et al. 2009). The overall consequences for the traditional maize cultivation system are nevertheless still unclear. It can be expected that certain traits, if conferring a selective advantage like insect resistance, might potentially increase in frequency. Despite unclear consequences for the local genetic diversity this may have socio-economic implications since GM impurities would close alternative agricultural development options, e.g. to grow organic crops. Because of the high natural biodiversity of landraces at the centre of origin of the species any contamination that puts the diversity at risk has to be avoided. Independent investigations on long-term implications for the diversity and the livelihood of the traditional farmers are required. In this context, the efficiency of the regulatory oversight regarding intended and unintended GM dispersal is also of concern.

Example 2: Inevitable Gene Escape in Africa

Small-scale subsistence maize cultivation cannot be controlled for seed purity.
In Africa, maize subsistence cultivation on very small plots is widespread and has a high importance for the livelihood of the rural population. Re-sowing of harvested crops, seed exchange among growers and cross-pollination between fields play an important role. In many regions, large numbers of small-scale farming predominate while also larger farms exist (Aheto et al. 2011; Viljoen & Chetty 2011). It can be expected that unintended dispersal of GM maize varieties (commercialised e.g. in South Africa) occurs. Enforced distance regulations facilitating trait segregation as well as efficient harvest purity controls are widely lacking. Successive dispersal to other regions is likely to proceed even across national borders, regardless of admission policies. Relevant undesirable effects have been reported from South Africa: Target pests became resistant against the toxins genetically engineered into maize (Kruger et al. 2012). The spatial proximity of different production regimes and deficits in the enforcement of a high (toxin) dose/refuge cultivation strategy (Tabashnik et al. 2004) may have played an additional role in this context.

In Africa, comprehensive investigations of transgene spread on a large scale are widely lacking and represent an open research task deserving high priority.

Fig. 1: Feral oilseed rape at a rail track distributing seeds (Bremen, Germany)

Exampe 3: Transgenic oilseed rape widely disperses

Transgenic oilseed rape has been found growing wild in Europe and Japan despite no commercial cultivation was allowed on a continental scale.

Transgenic oilseed rape survives in the wild and can hybridise with several related other species, some of them weedy, others cultivated (Menzel 2006). Transgenic varieties have been reported to have occurred very far from cultivation sites due to transport losses even across continents (Kawata et al. 2009; Schoenberger & d'Andrea 2012). In oilseed rape, transgenic herbicide resistance confers a selective advantage because typical habitats where the plants frequently grow are typical herbicide application sites (e.g. rail tracks, road margins, industrial areas, wasteland). Since gene flow is possible from oilseed rape to a number of wild and weedy relatives (OECD 1997; Halfhill et al. 2004), more systematic research is required concerning large-scale dispersal dynamics (Breckling et al. 2011). Transgenic individuals have already been found in the wild with new combinations of transgenes that were originally engineered into different varieties (Hall et al. 2000; Knispel et al. 2008). Thus, it is urgent to follow the ongoing spread and scientifically analyse short- and long term consequences.

Outlook

Evolutionary news that we will likely see in future are spread and re-combination of transgenes, giving rise to new properties. In a recent study, Wegier et al. (2011) have observed for the cultivation of cotton in Mexico that there is an ongoing molecular dynamic in terms of the dispersal of GM varieties into the centre of origin of the species following a metapopulation dynamic. Collapses of GM pest resistances, as upcoming in South Africa and recently also in the US (Gassmann et al. 2011) are likely and may confer the insight that the involved GM traits are not likely to hold much longer than the time span of patent protection. In case of a commercialisation of herbicide resistant grasses (e.g. *Agrostis stolonifera* for golf courses, Reichman et al. 2006) it could well be expected that we might witness the collapse of certain GM production forms based on direct seeding due to the proliferation of weedy GM grasses.

Overall, large-scale and long-term consequences of GM releases and cultivation require more research efforts. The currently available studies give hints that self-organised processes are taking place, however, the "what and where" is by far not sufficiently understood.

References

Acevedo, F., Huerta, E., Burgeff, C., Koleff, P., Sarukhán, J. (2011) Is transgenic maize what Mexico really needs? Nature Biotechnology 29: 23–24. doi:10.1038/nbt.1752.

Aheto, D.W., Reuter, H., Breckling, B. (2011) A modeling assessment of geneflow in smallholder agriculture in West Africa. Environ Sci Eur 23(9). doi:http://dx.doi.org/10.1186/2190-4715-23-9.

Breckling, B., Reuter, H., Middelhoff, U., Glemnitz, M., Wurbs, A., Schmidt, G., Schröder, W., Windhorst, W. (2011) Risk indication of genetically modified organisms (GMO): Modelling environmental exposure and dispersal across different scales: Oilseed rape in Northern Germany as an integrated case study. Ecological Indicators 11 (4): 936–941.

Chapela, I., Quist, D. (2001) Transgenic DNA introgressed into traditionalmaize landraces in Oaxaca, Mexico. Nature, 414(6863), 541–543.

Marvier, M., van Acker, R.C. (2005) Can crop transgenes be kept on Leash? Frontiers in Ecology and the Environment, 3(2), 93–100.

Ellstrand, N.C. (2003) Current knowledge of gene flow in plants: implications for transgene flow. Philosophical Transactions of the Royal Society of London. Series B: Biological Sciences, 358(1434): 1163–1170.

Gassmann, A.J., Petzold-Maxwell, J.L., Keweshan, R.S., Dunbar, M.W. (2011) Field-Evolved Resistance to Bt Maize by Western Corn Rootworm. PLoS One, 6(7), e22629. doi:10.1371/journal.pone.0022629.

Graef, F., Römbke, J., Binimelis, R., Myhr, A.I., Hilbeck, A., Breckling, B., Dalgaard, T., Stachow, U., Catacora-Vargas, G., Bøhn, T., Quist, D., Darvas, B., Dudel, G., Oehen, B., Meyer, H., Henle, K., Wynne, B., Metzger, M.J., Knäbe, S., Settele, J., Székács, A., Wurbs, A., Bernard, J., Murphy-Bokern, D., Buiatti, M., Giovannetti, M., Debeljak, M., Andersen, E., Paetz, A., Dzeroski, S., Tappeser, B., van Gestel, C.A.M., Wosniok, W., Séralini, G.E., Aslaksen, I., Pesch, R., Maly, S., Werner, A. (2012) A framework for a European network for a systematic environmental impact assessment of genetically modified organisms (GMO). BioRisk 7: 73–97. doi: 10.3897/biorisk.7.1969

Halfhill, M.D., Zhu, B., Warwick, S.I., Raymer, P.L., Millwood, R.J., Weissinger, A.K., Stewart, Jr. C.N. (2004) Hybridization and backcrossing between transgenic oilseed rape and two related weed species under field conditions. Environmental Biosafety Research, 3(2), 73–81., doi: 10.1051/ebr:2004007.

Hall, L., Topinka, K., Huffman, J., Davis, L., Good, A. (2000) Pollen flow between herbicide-resistant *Brassica napus* is the cause of multiple-resistant *B. napus* volunteers. Weed Science 48(6): 688–694. doi: 10.1614/0043-1745(2000)048[0688:PFBHRB]2.0.CO;2

Kawata, M., Murakami, K., Ishikawa, T. (2009) Dispersal and persistence of genetically modified oilseed rape around Japanese harbours. Environmental Science and Pollution Research, 16(2): 120–126. doi: 10.1007/ s11356-008-0074-4.

Knispel, A.L., McLachlan, S.M., Van Acker, R.C., Friesen, L.F. (2008) Gene Flow and Multiple Herbicide Resistance in Escaped Canola Populations. Weed Science 56(1): 72–80. doi: 10.1614/WS-07-097.1

Kruger, M., Van Rensburg, J.B.J., Van den Berg, J. (2012) Transgenic Bt maize: farmers' perceptions, refuge compliance and reports of stem borer resistance in South Africa. Journal of Applied Entomology 136 (1–2): 38–50, doi: 10.1111/j.1439-0418.2011.01616.x

Menzel, G. (2006) Verbreitungsdynamik und Auskreuzungspotenzial von *Brassica napus* L. (Raps) im Großraum Bremen. GCA-Verlag, Waabs, Dissertation University of Bremen

Ober, S. (2009) Risiken der Agrogentechnik untersuchen. 9-Punkte-Katalog für eine ökologische Risikoforschung.
http://www.boelw.de/uploads/media/Katalog_Sicherheitsforschung.pdf

OECD Organisation for Economic Co-Operation and Development (1997) Consensus Document on the Biology of *Brassica napus* L. (Oilseed rape). General Distribution. Series on Harmonization of Regulatory Oversight in Biotechnology No.7. OECD Environmental Health and Safety Publications (97)63, Paris,
http://search.oecd.org/officialdocuments/publicdisplaydocumentpdf/?cote=OCDE/GD%2897%2963&docLanguage=En (accessed 2012/10/9).

Piñeyro-Nelson, A., van Heerwaarden, J., Perales, H.R., Serratos-Hernández, J.A., Rangel, A., Hufford, M.B., Gepts, B., Garay-Arroyo, A., Rivera-Bustamante, R., Álvarez-Buylla, E.R. (2009) Transgenes in Mexican maize: molecular evidence and methodological considerations for GMO detection in landrace populations. Molecular Ecology 18(4):750–761.
doi: 10.1111/j.1365-294X.2008.03993.x

Reichman, J.R., Watrud, L.S., Lee, E.H., Burdick, C.A., Bollman, M.A., Storm, M.J., King, G.A., Mallory-Smith, C. (2006) Establishment of transgenic herbicide-resistant creeping bentgrass (*Agrostis stolonifera* L.) in nonagronomic habitats. Molecular Ecology 15 (13): 4243–4255 doi: 10.1111/j.1365-294X.2006.03072.x

Schmidt, G., Breckling, B. (2010) The Triffid case: A short résumé on the re-discovery of a deregistered GMO. In: Breckling, B. & Verhoeven, R. (2010) Implications of GM-Crop Cultivation at Large Spatial Scales. Theorie in der Ökologie 16. Frankfurt, Peter Lang. 79–81.

Schoenenberger, N., D'Andrea, L. (2012) Surveying the occurrence of subspontaneous glyphosate-tolerant genetically engineered *Brassica napus* L. (Brassicaceae) along Swiss railways. Environmental Sciences Europe 24:23, doi:10.1186/2190-4715-24-23

Tabashnik, B.E., Gould, F., Carrière, Y. (2004) Delaying evolution of insect resistance to transgenic crops by decreasing dominance and heritability. Journal of Evolutionary Biology 17 (4): 904–912, doi: 10.1111/j.1420-9101.2004.00695.x

Viljoen, C., Chetty, L. (2011) A case study of GM maize gene flow in South Africa. Environmental Sciences Europe, 23(1): 1–8. doi: 10.1186/2190-4715-23-8

Wegier, A., Pineyro-Nelson, A., Alarcón, J., Gálvez-Mariscal, A., Álvarez-Buylla, E.R., Pinero, D. (2011) Recent long-distance transgene flow into wild populations conforms to historical patterns of gene flow in cotton (*Gossypium hirsutum*) at its centre of origin. Molecular Ecology 20 (19): 4182–4194, doi: 10.1111/j.1365-294X.2011.05258.x

Chapter V

Transdisciplinary contributions:
Comments by stakeholders and administrators

Breckling, B. & Verhoeven, R. (2013) GM-Crop Cultivation – Ecological Effects on a Landscape Scale.
Theorie in der Ökologie 17. Frankfurt, Peter Lang.

Establishment of an European data centre for post market monitoring of GMO – what will be the best option?

Wiebke Züghart

Federal Agency for Nature Conservation, Bonn, Germany

Extended Abstract[1]

The need for a focal data centre for Post Market Monitoring (PMM) of genetically modified organisms is expressed widely by Member States and the European Commission (ENV/06/10 2006). The aim is to develop a common structure to
- collect and store PMM data (PMM reports),
- to interlink and manage data from different sources,
- enable comparability of data and / or analysed data sets,
- allow central data analysis,
- enable extrapolation to larger areas,
- allow extraction of overall trends,
- to link PMM data with environmental background data
  (e.g. climate change, land use change),
- facilitate transparency
  (e.g. assessments, decision making).

Provided that comparable environmental data are available across Europe, a common structure would facilitate aggregated analysis and evaluation for the purposes of PMM.

Though Post Market Monitoring is mandatory in the European Union, the incoming monitoring reports are evaluated only "by hand". No data management is yet implemented. Also relevant information and monitoring data generated in other fields of environmental assessment and monitoring are not yet used systematically for the purpose of PMM. Existing technical settings at national or European level are not yet sufficient to detect e.g. cumulative or unanticipated impacts of GMOs on the environment, which is a main task of PMM (BfN, FOEN, EEA 2011).

Different options on how to initiate an European data centre circulate among stakeholders, for instance the use of institutions such as EFSA, JRC or EEA or alternatively the development of a new information platform. Important criteria which have to be considered in this discussion are the existence of an infrastructure for environmental data

---

[1] A full paper was submitted to ESEU, Environmental Sciences Europe, Series: Implications of GMO-cultivation and monitoring.

storage, support and information flow, the host of public environmental data bases potentially relevant for PMM and the available expertise in analysing data of different sources. Considering these criteria, we suggest that the European Environment Agency would be the most suitable institution to connect GMO monitoring data with other environmental observations to draw coherent conclusions.

References:

ENV/06/10 (2006): Concepts for the coordination and harmonisation of monitoring data exchange regarding GM crops. European Commissions Working Group on Guidance notes supplementing Annex VII of Directive 2001/18/EC.

BfN, FOEN, EAA (2011): Monitoring of genetically modified organisms. A policy paper representing the views of the National Environmet Agencies in Austria and Switzerland and the Federal Agency for Nature Conservation in Germany. Vienna, REP-0305.

Official seed monitoring as a potential data source for GMO monitoring

Hans-Georg Starck

Ministry for Energy, Agriculture, the Environment and Rural Areas Schleswig-Holstein, Kiel, Germany

Abstract

The competent authorities in the EU member states are responsible for the enforcement of the regulations on GMO. In case of an unapproved deliberate release or placing on the market is discovered, the according legal actions must be taken. Legal basis is Art. 4, § 5 of the directive 2001/18/EC. In the year 2000, a number of illegal GMO imports of maize, oil seed rape and cotton seeds were detected in the EU. Since that time some (not all) member states have started a more or less systematic and continuous routine control of imported seeds to detect eventual GMO contaminations. Data of the last years from Germany but also from other EU member states show that seeds of some crops from particular origins have a permanent and significant risk to be contaminated. Thus, risk orientated seed monitoring is a strong approach to optimize GMO monitoring. These additional data may be of interest for the determination of a "baseline" before the deliberate release of new genetically modified varieties. It may help to specify background impurity rates which are important to know since routine testing targets event unspecific GM constructs.

Context

In 2002 a German Federal Working Group has developed a concept for the monitoring of genetically modified organisms (GMO) (UBA 2002). One main conclusion was that the ensuring of the purity of seeds and products is fundamental to protect the environment against an eventual uncontrollable dispersion of genetically modified plants. The subsequently established official GMO seed monitoring (inspection) is an already existing tool for a national general surveillance.

After a decade of official seed monitoring in Germany there are still some challenges which have to be met in order to assure and increase the efficiency of the programme:
- What has to be sampled? For an effective and risk orientated monitoring it is necessary to know, which imported seeds could be potentially contaminated. For that it is also necessary to have a complete overview of events worldwide and of any import into the EU.

- How to detect?
  It is essential to have appropriate detection methods and reference material available. The official laboratories must fulfil high standards both in the detection methods (in the case of a lawsuit) and in promptness.
- When are the samples to be collected?
  The best time for sampling is before the seeds enter the market. Otherwise it would be considerably more difficult to retrieve contaminated seed. The worst case is when contaminations are discovered after the seeds are already sown.
- Where are the samples to be taken?
  It is recommended to sample at places, which function as bottlenecks in the distribution chain. These are seed processing units or large distributors at an early stage of the distribution.
- The problem of traceability and retrieval still exists.
  In the case of a GMO contamination the whole chain from the seed company to distributors to retailers and to farmers has to be checked.

Seeds as a vector for GMO entries in the environment

Although there is still no European database for the results of the official seed monitoring (which should be a definite request to the European institutions), some data are existing. A survey from the European Union about seed contaminations in the years 2001 to 2006 has indicated that an average of 3.2 % of the tested seed lots had a GMO contamination (Hugo et al. 2007). Currently, data of the European Enforcement Project on Contained Use and Deliberate Release of GMO, a network of European official GMO inspectors, suggest that in some cases 3–7.5 % of tested sees lots (maize and rape seed) have GMO contaminations (oral note).

The existing data indicate a clear background emission of genetically modified plants (GMP) in seeds, especially in maize seeds from countries with a high proportion of GM-cultivation, currently or in the past (France, Hungary, Chile, United States, Canada). Therefore, a European-wide harmonised seed monitoring with harmonised detection limits and harmonised ratio of numbers of samples is imperative both for the compliance with the existing co-existence regulations and for the general surveillance of GMP in the environment.

The European institutions have to collect and to publish all incidents. Furthermore the official seed monitoring needs more information about "risky" seeds, i.e. which countries have a high ratio of deliberate releases and seed multiplication of GMP, which events are used, how high the capability for outcrossing or transgene escape and volunteers is. Furthermore, a survey of the particular seed trade volume is required. As a first step it is recommended to extend the sampling to more different crops for instance maize, rapeseed, sugar beet, potato, mustard, flax seed, radish, alfalfa, soybeans, and bent grass.

References

Hugo, S., Macarthur, R., Dixon, J., Murray, A. et al. (2007) Adventitious traces of genetically modified seeds in conventional seed lots: Current situation in the member states. Research tender ENV.B.3/ETU/2006/0106R. Central Science Laboratory, Sand Hutton.
http://www.gm-inspectorate.gov.uk/reportsPublications/documents/EUseeds_final_081007.pdf

UBA (2002) Entwurf eines Konzepts für das Monitoring von gentechnisch veränderten Organismen (GVO). http://www.bfn.de/fileadmin/MDB/documents/BLAGKonzept02-12-10.pdf

Impact of GMOs on the beekeeping sector: Still neglected in research, widely ignored in regulation, untapped potential for monitoring

Walter Haefeker

European Professional Beekeepers Association, Seeshaupt, Germany

Abstract

Beekeepers around the world have pointed out considerable deficits in assessing the implications of GMO cultivation for beekeeping and in maintaining the conditions for GMO-free honey production. Since the foraging range of bees encompasses several kilometres, bees and bee products are uniquely susceptible to exposure from GMO crops. There are clear regulatory requirements to assure the possibility for the production of GMO free honey as well as the need for GMO registers and proper monitoring.

It is proposed here to considerably expand the use of bee keeping to contribute to GMO monitoring. A global database of gene flow could easily be established and maintained in partnership between the scientific community and the honey sector, which already analyzes honey on a regular basis for GMO content and has a vested interest in maps indicating high or low risk of contamination.

Introduction

Since honey bees forage across a distance of several kilometres, honey production is at a high risk of containing unintended traces of GMO if modified plants are grown under the current regulatory regime in the EU. Only a relatively small distance to conventional crops of the same species must be kept, whereas the requirements of gm free honey production are not taken into consideration. To be able to serve the customers preferences for GM free honey, bee keepers actively demand to safeguard the potential of GM free honey production on the landscape scale. This is particularly relevant for beekeepers in the EU and beekeepers outside of the EU who export to Europe, where the customers are particularly sensitive to the GMO issue and have been promised freedom of choice, coexistence and zero-tolerance for all food by the European Commission (EC 2003).

By labelling the product consumers are supposed to be enabled to choose between GM or non-GM food. The production of non-GM food is to be protected through appropriate coexistence measures implemented in the member states. Finally, only events that have been tested and approved for human consumption may be placed on the EU market.

Open systems facilitate complex impact

Honey bees cover large foraging areas of several square kilometres. When bees collect nectar, pollen, honeydew, resin and water, they do not distinguish between conventional and genetically modified plants (Haefeker 2008). Such an extremely open production system like bee keeping raises a very complex set of problems for an appropriate regulation. Regulators, tend to think in terms of clearly delineated systems with defined interfaces. Coexistence measures of beekeeping GM crop cultivation clearly would require not meters but kilometres of separation distance, putting large areas of farm land out of reach for GMO cultivation. It is quite obvious that GM crop cultivation and GM free honey production are hard to reconcile in the same region. It is equally obvious that the current regulatory norms do not offer a solution for the existing conflicts but leave the interests of honey producers out of consideration.

Records show, that the EU Commission as early as 2004 began to try to circumvent it's own co-existence principles. The Standing Committee on the Food Chain and Animal Health (2004) discussed ways to effectively remove honey from GMO regulations. This left beekeepers in a legal limbo. Regulators in EU member states pretended that there was no requirement to include beekeeping in their coexistence schemes, even though there was no document with legal force exempting honey or any other bee products from GMO regulations governing the presence of GMOs in food.

European Court of Justice ruling

In 2011 the beekeepers succeeded to get the European Court of Justice to rule on this matter (BMELV 2011; ECJ 2011). The Court clearly agreed with the beekeepers, that honey cannot remain outside of the GMO regulatory framework. The Court could have decided, that the Regulation (EC) No 1829/2003 (EC 2003) and others have to be changed or amended to rectify this problem and set appropriate norms. Instead, the Court found a way to remain within the existing regulation, by deciding, that (GM-) pollen in honey has to be treated like a food ingredient pursuant to Regulation (EC) No 1829/2003 and thus undergoes the 0.9% labelling requirement. Apparently, major problems remain unsolved. It is wrong to assume that the court decision has relevant effect on the regulation of honey not containing GMOs. Nothing in the ruling requires any regulatory body to change how GMO-free honey is labelled or how GM free hone production can be secured.

Global implications

Thus, any country with honey production for the EU market needs to implement tight controls and measures for the cultivation of GMOs in order to minimize the impact on the beekeeping sector. A carefully monitored public register of all GMO cultivation including research plots is an essential element of any proper regulatory regime. Failure to provide accurate information about where the risk of contamination exists results in prohibitively high analysis cost for GMO-free production of honey and other bee products.

There are examples of public registers able to provide useful information to beekeepers. German GMO legislation has been providing for a public register of GMO fields. After the court ruling the German Ministry of Agriculture published a Q&A document stating that beekeepers may rely on this register when declaring the GMO status of their honey. The German organization for GMO free labelling VLOG, also allows beekeepers to rely on the "Standortregister" (VLOG 2011).

As a major honey exporting country with customers in the EU, Chile was affected by the verdict of the European court of justice. Chile's Agriculture and Livestock Service (SAG) responded quickly in December of 2011 by establishing a computer system where beekeepers can enter the coordinates of their apiaries to receive data on the proximity of genetically modified crops (FreshFoodPortal 2011). In addition, the Chilean "Transparency Council" ruled on March 23rd, 2012 in favour of the rights of citizens including beekeepers to know the location of GMO fields (GMWatch 2012).

Currently there are very few EU member states with co-existence measures designed to protect the GMO free production of honey and other bee products. GMO registers and monitoring programs are clearly needed by the beekeeping sector to ensure the information in GMO registers is accurate.

Beekeeping sector's contribution to GMO monitoring efforts

The potential exists for the bee keeping to contribute to GMO monitoring. Since the bee's activity range can be the same order of magnitude as pollen transport by wind for many wind-pollinated crops and bees collect pollen independent from nectar, the presence of GMO crops can be detected on the regional scale. This holds true even if the crops are not directly visited by bees, because wind carried pollen attaches to sticky honey dew, which is released by aphids and frequently collected by bees (Hofmann et al. 2005). Routine honey sampling could be established as an excellent tool for large scale monitoring for GMO presence-absence assessment. A global database of gene flow could be created and maintained in partnership between the scientific community and the honey sector, which already analyzes honey on a regular basis for GMO content and has a vested interest in maps indicating high or low risk of contamination.

Conclusion

The perspective of the beekeeping sector can no longer be neglected in GMO risk research and risk management. There is a clear requirement to fill research gaps, improve regulation and monitoring while at the same time realizing the potential of bees for mapping global GMO gene flow.

References

BMELV (Bundesministerium für Ernährung, Landwirtschaft und Verbraucherschutz) (2011) Fragen und Antworten zum "Honig-Urteil" des Europäischen Gerichtshofs, BMELV, September 2011.
www.bmelv.de/SharedDocs/Standardartikel/Landwirtschaft/Pflanze/GrueneGentechnik/FAQGentechnikUrteilSeptember2011.html

ECJ (Court of Justice of the European Union) (2011) ECJ Ruling1;
http://curia.europa.eu/jurisp/cgi-bin/gettext.pl?where=&lang=en&num=79889093C19090442&doc=T&ouvert=T&seance=ARRET

EC (2003) Commission of the European Communities (2003) Commission recommendation on guidelines for the development of national strategies and best practices to ensure the co-existence of genetically modified crops with conventional and organic farming. 23. 6. 2003.
http://ec.europa.eu/agriculture/publi/reports/coexistence2/guide_en.pdf

FreshFoodPortal (2011) New tools for Chilean beekepers to meet demanding European GM standards. December 9th, 2011.
http://www.freshfruitportal.com/2011/12/09/new-tools-for-chilean-beekepers-to-meet-demanding-european-gm-standards

GMWatch (2012) Chilean Citizens win right to know GMO cultures location, GM-Watch, 29. 3. 2012. http://gmwatch.org/component/content/article/13803

Haefeker, W. (2008) Co-existence of GM-crops with beekeeping – impact of GM-crops on the supply chain for honey and other bee products. In: Breckling, B., Verhoeven, R.: Implications of GM-Crop Cultivation at Large Spatial Scales. Theorie in der Ökologie 14. Frankfurt, Peter Lang: 131–133.

Hofmann, F., Schlechtriemen, U., Wosniok, W., Foth, M. (2005) GVO-Pollenmonitoring. Technische und biologische Pollenakkumulatoren und PCR-Screening für ein Monitoring von gentechnisch veränderten Organismen. BfN-Skripten 139.
http://www.bfn.de/fileadmin/MDB/documents/skript139.pdf

Standing Committee on the Food Chain and Animal Health (2004) Summary record of the 2nd meeting, 23rd June 2004.
http://ec.europa.eu/food/plant/standing_committees/sc_modif_genet/docs/summary02_en.pdf

VLOG (2011) Verband Lebensmittel ohne Gentechnik e.V.: Kriterien für die „Ohne Gentechnik" Auslobung von Honig und anderen Imkererzeugnissen, Stand 12/2011.
http://www.ohnegentechnik.org/fileadmin/ohne-gentechnik/Das_Siegel/OG_Kritierien_Honig_und_Erzeugererklaerung_111220.pdf

Breckling, B. & Verhoeven, R. (2013) GM-Crop Cultivation – Ecological Effects on a Landscape Scale.
Theorie in der Ökologie 17. Frankfurt, Peter Lang.

Deficits in research funding for analysis of health and environmental risks of GM plants – the example of Germany

Martha Mertens

Institute for Biodiversity Network, Regensburg, Germany

Abstract

Public funding of research is essential to guarantee that the public interests in the safe application of biotechnology are not overrun by private economic interests. To achieve this, industry-independent research and expertise has to extend to the identification of health and environmental effects of transgenic plants that are not the major focus of product development. In Germany, many programs have been launched to support biotechnology, but research programs into the risks and impacts of transgenic plants are underrepresented.

Introduction

More than 15 years after the introduction of genetically modified organisms (GMO) into agriculture and the food chain the risk issues are still hotly debated. To guarantee that the public interests in the safe handling of biotechnology are not overrun by the economic interests of developers and adopters of GMOs, public funding of research into risks connected to the use of GMOs is essential. While industry-independent research and expertise has to be on the same technological level as industrial research, it also has to extend to fields not addressed in the product development such as long-term and large-scale effects and socio-economic impacts of GMOs.

Funding programs for (plant) biotechnology in Germany

From 2003 to 2011, the main funding agency, the Federal Ministry for Education and Research (Bundesministerium für Bildung und Forschung BMBF), spent a total of € 2.86 billion on biotechnology including drug development (http://foerderportal.bund.de). Several funding programs were installed, directed to young technology companies (High-Tech Start-up Fund), small- and medium-sized companies (Bio Chance), young scientists (BioFuture) and to regions active in "modern biotechnology" (BioRegio and BioProfiles). From 2007 to 2011 the program GABI (genome analysis in the biological system plant) received > € 60 million (GABI 2012).

Deficits in research funding for analysis of health and environmental risks ...

In adopting the national BioEconomy 2030 research strategy, the German government pursues "the vision of a sustainable, biobased economy that carries with it the promise of global food supplies that are both ample and healthy and of high-quality products from renewable resources" and declares "great potential is seen in plant genetic engineering" (Van Liempt 2011). In the period 2011–2014, funding of plant biotechnology amounts to about € 43.9 million (http://foerderportal.bund.de).

Private companies, among them big players such as BASF Plant Science, Bayer Crop Science and KWS, have been funded too. In the period 2008 to 2011 they received € 5 million for developing genetically modified plants (GM plants), e.g. oilseed rape, barley, wheat, sugar beet, rice, and chicoree (BR 2009) – without any relation to biosafety research.

Programs for GM plants risk research

As the ministry put it, it perceives a need to be able to assess the potential results and consequences of the technology. Funding independent safety research into GM crops is therefore "a part of acting responsibly in the interests of human, animal, and environmental safety" (Van Liempt 2012).

Within the last 25 years the BMBF spent more than € 100 million on about 300 projects in biosafety research. 120 projects were directly related to risk assessment of GM plants. In 2008–2011 about € 12 million were allocated under this headline (BR 2009). More than 80 % of this money went into projects such as:
- improvement of biosafety of GM plants, e.g. male sterile systems or gene targeting (€ 1.8 million)
- confinement strategies for oilseed rape (€ 1.12 million)
- transformation of plastids in oilseed rape and maize (€ 1.17 million)
- research accompanying release of Bt maize (€ 2.63 million)
- biosafety of transgenic trees (€ 0.55 million)
- authorisation and monitoring of GM plants (€ 2.82 million, > 75 % of the amount went to the private sector).

In some of the projects new GM plants were developed (e.g. maize, oilseed rape, potato, and apple). Communication of biosafety research, performed by a private company, has been heavily financed too: from 2005–2009 nearly € 2 million were allocated to this end and running the website www.gmo-safety.eu cost € 250.000 a year (BR 2009).

BMBF: Investment in biosafety research has been worthwhile

One of the bigger projects funded in the past was the Bt maize project with Monsanto lines expressing single and multiple Cry proteins. The main aim was to study effects on non-target organisms, including honey bees, and to collect data on Cry proteins in soil and their effects on soil microflora. Scientists involved in the project concluded: "GM Bt maize is as safe as conventional maize." They further concluded that Bt maize is better for protecting species diversity in the fields and its cultivation could help prevent soil erosion and conserve soil fertility. They also suggested that other countries and international organisations might profit from the biosafety research done in Germany (BMBF 2012).

According to BMBF, none of the GM plants investigated so far have been found to have harmful effects on the environment. "The BMBF intends to continue to promote plant biotechnology, particularly in view of the proven environmental safety of GM plants, and is advocating freedom of research and openness towards new technologies" (BMBF 2012).

Few examples of critical research

As shown above, to a large extent public resources are used to support GM product development and communication of the benefits of GM plants. Only a few projects that address large-scale and long-term consequences of GMOs use have been funded. One example is the GeneRisk project (2005–2010) dealing with the ecological, legal and economic analysis of coexistence of agriculture with and without GM plants (http://www.sozial-oekologische-forschung.org/de/692.php). The project focused on analysis of systemic risks linked to GM plant use, such as impacts on ecosystems, landscapes, agricultural practice, and socio-economic and legal systems.

Also, the German government provides no public funding for research on herbicide resistant (HR) crops (BR 2009), despite the fact that more than 80 % of the GM crops worldwide carry herbicide resistance (HR) genes. 37 HR crop lines are authorized in the EU for food and feed (http://ec.europa.eu/food/dyna/gm_register/index_en.cfm) and more than 10 HR crops await the approval for cultivation (www.gmo-compass.org). New data indicate that the herbicides used with these HR crops are less benign than proposed, that they are toxic to humans, animals, soil life, and aquatic organisms and that their use rate has increased tremendously. There is a lack of data about their residues in food, feed, drinking water and the environment.

Other research areas not fully addressed by public funding include:
- long-term and large-scale impacts of GM plants on human and animal health and the environment (including non-target organisms and pollinators),
- stability of gene expression in varying environments,
- standardization of test systems,
- impacts of Bt and HR crops: e.g. toxicity, resistance evolution in weeds and pests, pesticide use,
- interactions of genes and traits in stacked trait GM plants,
- coexistence issues: gene spread, occurrence of related plants,
- socio-economic effects: costs imposed on GM-free agriculture and food production.

In conclusion, there is still a striking imbalance in German public funding of GMO risk research. Public resources are spent to a large extent on development of transgenic plants and even private companies, including the big players, are supported. When it comes to programs on biosafety research, then again a considerable amount of money is diverted to development of GM plants and the communication of their perceived benefits. For future funding periods, the German government should be requested to support truly independent research into health and environmental risks of GM plants.

References

BMBF Bundesministerium für Bildung und Forschung (2012) Biosafety research conclusions. Genetically modified Bt maize is safe.
    http://www.gmo-safety.eu/news/1388.igw-genetically-modified-biosafety-research.html.
BR Bundesregierung (2009) Antwort auf Kleine Anfrage von Bündnis90/Die Grünen: Risikoforschung und -prüfung bei gentechnisch veränderten Pflanzen, Drucksache 16/12969.
    http://dipbt.bundestag.de/dip21/btd/16/129/1612969.pdf.
GABI Genomanalyse im biologischen System Pflanze (2012) Alle Projekte.
    http://gabi.de/projekte-alle-projekte.php.
Van Liempt, H. (2011) Results of BMBF-funded biosafety research into GM crops.
    http://www.gmo-safety.eu/news/1311.biosafety-research-genetically-modified-crops.html.

Research Policy & Independent Risk Research – Draft demands by civil society organisations for German parliamentary elections in 2013

Christoph Then

Testbiotech, München, Germany

Introduction

Current European and German research policy is criticised by many civil society organisations. Drastic problems such as climate change, decreasing resources and global food supply are fueling the discussion. Uneasiness about a research policy that is mostly driven by economic interests is increasing.

It's time for a change: In 2011 about 100 civil society organisations published a call for a substantial change of the common strategic framework for the funding of European research and innovation programmes. In its 2001 annual report , the German Advisory Council on Global Change (WBGU) called for a "social contract for sustainability". Most recently, in February 2012, the German organisation, Friends of the Earth (FOE/BUND) published a report titled "Sustainable Science", which summarises several aspects of the current discussions on this subject.

One of the most important demands of organisations such as FOE (Germany) or the Nature and Biodiversity Conservation Union (NABU) is a multimillion sustainability programme, the so called "Nachhaltigkeitsmilliarde", a yearly fund to drive the necessary transformation of science to secure a real change of policy in the fields of energy, agriculture, mobility and sustainable urban development. Furthermore, calls are being made for new structural instruments to enable more participation, transparency and democratic control by society and thereby put an end to the current fixation on technology, competitiveness and economic growth.

Similar demands are raised in a letter to EU Commissioner, Barroso, which was signed by about 100 civil society organisations in 2011:

- Research that will make Europe (and the world) an environmentally sustainable, healthy and peaceful place to live must now be prioritised over and above research that delivers marketable technologies. We, the undersigned civil society and scientific organisations, think that another research and innovation policy is not only possible but urgently necessary in order to respond to the challenges our societies are facing. We call on the EU Institutions to take steps to.

- Overcome the myth that only highly complex and cost intensive technologies can create sustainability, employment and well-being, and focus on tangible solutions to environmental, economic and societal challenges instead;
- Ensure that the concept of innovation includes locally adapted and social forms of innovation as well as technological development, and facilitate cooperation and knowledge exchange between civil society organisations and academia in order to realise the innovative potential of the non-profit sector;
- Establish a democratic, participatory and accountable decision-making process for research funding allocation, free from conflicts of interest and industry dominance, and enable civil society to play a full part in both setting the EU research agenda and participating in all EU research programmes;
- Ensure that all experts advising EU research policy-makers are appointed in a transparent manner to provide impartial and independent expertise, free from conflicts of interests; replace industry-dominated advisory groups and technology platforms with bodies that provide a balanced representation of views and stakeholders;
- Ensure that publicly funded research benefits wider society by systematically requiring equitable access licensing and encouraging open source access policies in the next Common Strategic Framework."

The aim of the signatories of this letter is the initiation of a much more sustainable research policy in Germany and the European Union, beginning with the German parliamentary elections in 2013. To foster public debate, we support these "electoral benchmarks", directed at all the participating parties of the 2013 German parliamentary elections.

In addition to the demands presented above, we also see the need to build up more expertise in the fields of new, risky and high tech technologies, which is independent from the interests of industry. Similarly, as was the case prior to the recent financial crisis, only a few experts are investigating the long-term implications in great detail that may be associated with new developments in the fields of nanotechnology and biotechnology as well as areas such as energy, mobility and IT technologies. Many of the relevant experts have close ties with industry (for example via funding of their research projects) or are working for government authorities, which in many cases are more interested in enhancing economic growth and competitiveness than fostering a broader public debate. The establishment of a broad and well-founded counter expertise in the field of risky technologies, which is also represented on an institutional level is essential.

General political goals, we are aiming for:

- "Public money for public goods", the priority for publicly funded research policy should be the benefit of general society;

- More participation by civil society organisations in planning and implementing research policy;
- Much more transparent research policy frameworks should be developed, which are not| predominantly driven by economical interests;
- Enhancing independent counter expertise, especially in technological areas associated with higher risks for the environment and health;
- More scope for new, unconventional and controversial debates and views within the scientific community;
- Innovations in the fields of social and cultural sciences should be regarded as being equally important as technological innovations.

Specific "electoral benchmarks"

1. Participatory Research Policy & Independent Research Council
   We call for the establishment of an Independent Research Council with representatives from civil society (such as environmental and consumer organisations) and the scientific community. This council shall have a say and decide on research funding in areas such as nutrition, agriculture, energy, mobility and public health (the "multimillion sustainability programme"). Further this council shall consult the government regarding the development of the general strategic framework in terms of research and innovation funding.

2. Multimillion Sustainability Programme
   In order to warrant sufficient funding for a transformation of science and innovation and a real change of policy in the fields of energy, agriculture, mobility and sustainable urban development, we are calling for a yearly multimillion sustainability fund . The Independent Research Council shall decide about the distribution of this budget.

3. Establishment of independent risk research and counter expertise
   A broad range of independent counter expertise in specific technological areas that have a high risk potential or are subject to controversial discussions (such as biotechnology, nanotechnology, mobility, energy, IT technologies) shall be established. New research programmes and new structures are needed and should be interconnected with already existing institutions. Companies which invest in risky technologies shall contribute to a Fund for Risk Research. The Independent Research Council shall decide how to use funds to establish and develop an independent risk research and counter expertise.

4. Fostering transparency / Clearinghouse mechanisms
   For questions from the interested public, a so called Clearinghouse shall be established, which can provide relevant information independently and comprehensively, without bureaucracy, about public funding for research and innovation. This shall enable interested members of the public to interfere at the

early stages of decision-making processes with sufficient knowledge and education of the relevant subjects.

This text was elaborated at a workshop of civil society in Berlin on 22 of March 2012 (http://www.nabu.de/themen/umweltpolitik/nachhaltigeentwicklung/14749.html).

Editing team: Martha Mertens (FOE Germany, BUND), Hartmut Meyer (ENSSER), Claudia Neubauer (Foundation Sciences Citoyennes), Steffi Ober (NABU), Christoph Then (Testbiotech).

Breckling, B. & Verhoeven, R. (2013) GM-Crop Cultivation – Ecological Effects on a Landscape Scale. Theorie in der Ökologie 17. Frankfurt, Peter Lang.

A publication forum for GMO research, initiated by the GMLS conference:
Implications of Cultivation and Monitoring of Genetically Modified Organisms
Series, published in Environmental Sciences Europe, a SpringerOpen Journal

Gunther Schmidt & Broder Breckling

Chair of Landscape Ecology, University of Vechta, Germany

The GMLS conferences were initiated in the course of social ecological research in Germany, basing on the conviction that molecular biological expertise solely is insufficient to analyse and appraise the ecological, economic and social effects of the use of genetically modified organisms in open environments, in particular in agriculture.

Supported by the conference discussions, the organising institutions realised relevant research deficits with regard to large-scale processes regarding GMO: effects on farming systems, on ecosystems and landscapes, on the social context, the rural population, as well as on food supply, consumers, and their freedom of choice. Molecular biology has the methodological expertise to create GMO but apparently has not the methodological repertoire to follow and evaluate all relevant higher level interactions in the context of the ecological economic and social systems.[1]

To improve the accessibility and discussion of scientific findings on these organisation levels, the GMLS contributions were made available on the conference website[2] for free download supplementary to the printed version of the proceedings. In addition, the organisers achieved to establish a focal GMO series[3] in an independent international journal addressing environmental issues. Environmental Sciences Europe (ESEU) is an international, peer-reviewed open access journal, focusing primarily (but not exclusively) on Europe, with a broad scope covering all aspects of environmental sciences, including the main topics of regulation. The target groups are environmental scientists from academia, but also industry and administrative bodies as well as regulators / decision-makers, both at the regional and European level, and beyond. This perfectly meets also the audience for which GMLS works.

---

[1] Breckling B, Schmidt S, Schöder W (Eds) 2012: GeneRisk. Systemische Risiken der Gentechnik: Analyse von Umweltwirkungen gentechnisch veränderter Organismen in der Landwirtschaft (GeneRisk systemic risks of genetic engineering: analysis of environmental effects of genetically modified organisms in agriculture, in German). Springer Heidelberg, 318 p.
[2] www.gmls.eu
[3] www.enveurope.com/series/GMO_cultivation

Since the start of the series in 2011, more than 20 contributions were published originating from GMLS keynote lectures together with additional papers dealing with related research on GMO. The download rates that can be used to measure the attention an article received. The recorded downloads largely exceeded the expectations of the journal's editors as well as the series' editors indicating that there is a high demand on scientific findings addressing ecological, environmental and social implications and controversies on GMO. By November 2012 the total number of downloads was about 163,000, whereas the most accessed contribution was downloaded more than 80,000 times, and another 5 contributions had more than 5,000 accessions within 1 ½ years, which is still very high for scientific journal contributions. Looking at the daily download rates, on average, articles from the entire series were downloaded 250 times per day, varying from 360 to 3 times per article (Fig. 1).

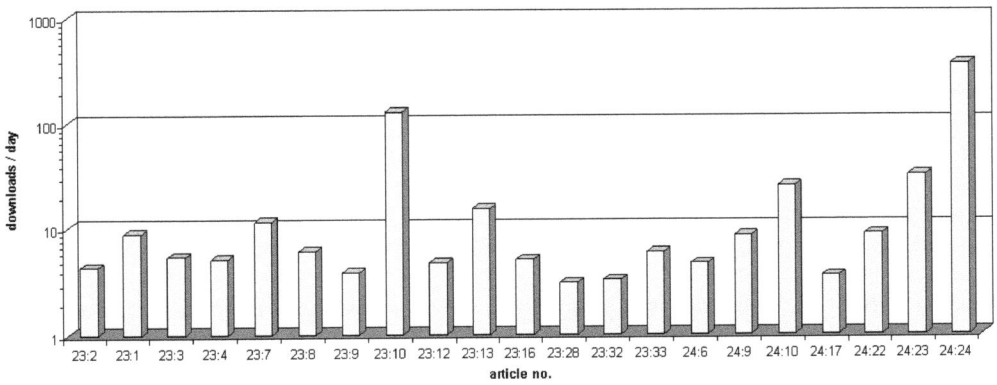

Fig. 1: Download rates per day for the articles published in the GMO series of ESEU (cut-off date: 21.11.2012)

Completing the third GMLS proceedings volume, we take the opportunity to give an overview on the contributions so far, and hope that this encouraging development will help to broaden the range of investigations – and also help to direct research attention (last not least research funding) towards large-scale interactions.

In fact, numerous open questions remain: The research group of Andres Carrasco attracted attention with their studies on the effects of GMO-based pesticide use in South America for human health. Consequences of large-scale application practice there are by far insufficiently reflected and analysed by scientific assessment.[4] Aheto et al. (2011) and Viljoen et al. (2011) pointed to the wide range of open questions posed by GMO releases in Africa, the same was addressed by Acevedo for Mexico.[5] GM oilseed rape

---

4 López, S.L., Aiassa, D., Benítez-Leite, S., Lajmanovich, R., Mañas, F., Poletta, G., Sánchez, N., Simoniello, M.F., Carrasco, A.E. (2012) Pesticides Used in South American GMO-Based Agriculture: A Review of Their Effects on Humans and Animal Models. Advances in Molecular Toxicology 6, 41-75.
5 Acevedo Gasman, F. (2010) Monitoring maize diversity in Mexico for decision making:
http://www.mapserver.uni-vechta.de/generisk/gmls2010/beitraege/GMLS2_Acevedo.pdf

dispersal far away from cultivation was reported by Kawata et al. (2011) for Japan and very recently also by Schönenberger & D'Andrea (2012) for Switzerland. Benbrook (2012) corroborated that GMO cultivation in the US largely increased pesticide use and environmental concerns.

These examples together with issues raised by other authors in the series give evidence that there is a great demand for interdisciplinary expertise from ecology, economy, social science, medicine, and others when assessing possible impacts not only in molecular contexts and business implications but also to illuminate the other relevant scientific domains.

Available contributions of the ESEU-Series on Implications of Cultivation and Monitoring of Genetically Modified Organisms

Schmidt G., Schröder W. (2011) Editorial: Implications of GMO cultivation and monitoring-series. Environmental Sciences Europe 23:2.

Taube F., Krawinkel M., Susenbeth A., Theobald W. (2011) The booklet "Genetically modified crops" published from the German Research Foundation, does not meet the given claim. Environmental Sciences Europe 23:1.

Pavone V., Goven J., Guarino R. (2011) From risk assessment to in-context trajectory evaluation – GMOs and their social implications. Environmental Sciences Europe 23:3.

Kleppin L., Schmidt G., Schröder W. (2011) Cultivation of GMO in Germany: support of monitoring and coexistence issues by WebGIS technology. Environmental Sciences Europe 23:4.

Meyer H. (2011) Systemic risks of genetically modified crops: the need for new approaches to risk assessment. Environmental Sciences Europe 23:7

Viljoen C., Chetty L. (2011) A case study of GM maize gene flow in South Africa. Environmental Sciences Europe 23:8.

Aheto D., Reuter H., Breckling B. (2011) A modeling assessment of geneflow in smallholder agriculture in West Africa. Environmental Sciences Europe 23:9.

Séralini G.E., Mesnage R., Clair E., Gress S., De Vendomois J., Cellier D. (2011) Genetically modified crops safety assessments: present limits and possible improve-ments. Environmental Sciences Europe 23:10.

Pascher K., Moser D., Dullinger S., Sachslehner S., Gros P., Sauberer N., Traxler A., Grabherr G., Frank T. (2011) Setup, efforts and practical experiences of a monitoring program for genetically modified plants – an Austrian case study for oilseed rape and maize. Environmental Sciences Europe 23:12.

Broer I., Jung C., Ordon F., Qaim M., Reinhold-Hurek B., Sonnewald U., von Tiedemann A. (2011) Response to the criticism by Taube et al. in ESE 23:1, 2011, on the booklet "Green Genetic Engineering" published by the German Research Foundation (DFG). Environmental Sciences Europe 23:16.

Winter G., Stoppe-Ramadan S. (2011) European Union and German law on co-existence: Individualisation of a systemic problem. Environmental Sciences Europe 23:28.

Hilbeck A., Meier M., Römbke J., Jänsch S., Teichmann H., Tappeser B. (2011) Environmental risk assessment of genetically modified plants – concepts and controversies. Environmental Sciences Europe 23:13.

von Kries C., Winter G. (2011) Legal implications of the step-by-step principle. Environmental Sciences Europe 23:32.

Dolezel M., Miklau M., Hilbeck A., Otto M., Eckerstorfer M., Heisenberger A., Tappeser B., Gaugitsch H. (2011) Scrutinizing the current practice of the environmental risk assessment of GM maize applications for cultivation in the EU. Environmental Sciences Europe 23:33.

Schröder W., Schmidt G. (2012) Overview of principles and implementations to deal with spatial issues in monitoring environmental effects of genetically modified organisms. Environmental Sciences Europe 24:6.

Hilbeck A., Meier M., Trtikova M. (2012) Underlying reasons of the controversy over adverse effects of Bt toxins on lady beetle and lacewing larvae. Environmental Sciences Europe. 24:9.

Hilbeck A., McMillan J.M., Meier M., Humbel A., Schläpfer-Miller J., Trtikova M. (2012) A controversy re-visited: Is the coccinellid Adalia bipunctata adversely affected by Bt toxins? Environmental Sciences Europe 24:10.

Wurbs A., Bethwell C., Stachow U. (2012) Assessment of regional capabilities for agricultural coexistence with genetically modified maize. Environmental Sciences Europe 24:17.

Bøhn T., Primicerio R., Traavik T. (2012) The German ban on GM maize MON810: scientifically justified or unjustified? Environmental Sciences Europe 24:22.

Schönenberger N., D'Andrea L. (2012) Surveying the occurrence of subspontaneous glyphosate-tolerant genetically engineered Brassica napus L. (Brassicaceae) along Swiss railways. Environmental Sciences Europe 24:23.

Benbrook C.M. (2012) Impacts of genetically engineered crops on pesticide use in the U.S. – the first sixteen years. Environmental Sciences Europe 24:24

Theorie in der Ökologie

Herausgegeben von Broder Breckling

Band 1 Broder Breckling / Felix Müller (Hrsg.): Der Ökologische Risikobegriff. Beiträge zu einer Tagung des Arbeitskreises "Theorie" in der Gesellschaft für Ökologie vom 4.-6. März 1998 im Landeskulturzentrum Salzau. 2000.

Band 2 Kurt Jax (Hrsg.): Funktionsbegriff und Unsicherheit in der Ökologie. Beiträge zu einer Tagung des Arbeitskreises "Theorie" in der Gesellschaft für Ökologie vom 10. bis 12. März 1999 im Heinrich-Fabri-Institut der Universität Tübingen in Blaubeuren. 2000.

Band 3 Hauke Reuter: Individuum und Umwelt. Wechselwirkungen und Rückkopplungsprozesse in individuenbasierten tierökologischen Modellen. 2001.

Band 4 Fred Jopp / Gerd Weigmann (Hrsg.): Rolle und Bedeutung von Modellen für den ökologischen Erkenntnisprozeß. 2001.

Band 5 Kurt Jax: Die Einheiten der Ökologie. Analyse, Methodenentwicklung und Anwendung in Ökologie und Naturschutz. 2002.

Band 6 Franz Hölker (ed.): Scales, Hierarchies and Emergent Properties in Ecological Models. 2002.

Band 7 Achim Lotz / Johannes Gnädinger (Hrsg.): Wie kommt die Ökologie zu ihren Gegenständen? Gegenstandskonstitution und Modellierung in den ökologischen Wissenschaften. Beiträge zur Jahrestagung des Arbeitskreises Theorie in der Gesellschaft für Ökologie vom 21.-23. Februar 2001 im Kardinal-Döpfner-Haus Freising (Bayern). 2002.

Band 8 Katrin S. Romahn: Rationalität von Werturteilen im Naturschutz. 2003.

Band 9 Hauke Reuter / Broder Breckling / Arend Mittwollen (Hrsg.): Gene, Bits und Ökosysteme. Implikationen neuer Technologien für die ökologische Theorie. 2003.

Band 10 Thomas Potthast (Hrsg.): Ökologische Schäden. Begriffliche, methodologische und ethische Aspekte. 2004.

Band 11 Angela Weil: Das Modell „Organismus" in der Ökologie. Möglichkeiten und Grenzen der Beschreibung synökologischer Einheiten. 2005.

Band 12 Fred Jopp / Silvia Pieper (Hrsg.): Bodenzoologie und Ökologie. 30 Jahre Umweltforschung an der Freien Universität Berlin. 2008.

Band 13 Boris Schröder / Hauke Reuter / Björn Reineking (eds.): Multiple Scales in Ecology. 2007.

Band 14 Broder Breckling / Hauke Reuter / Richard Verhoeven (eds.): Implications of GM-Crop Cultivation at Large Spatial Scales. Proceedings of the GMLS-Conference 2008 in Bremen. 2008.

Band 15 Denis Worlanyo Aheto: Implication Analysis for Biotechnology Regulation and Management in Africa. Baseline Studies for Assessment of Potential Effects of Genetically Modified Maize (Zea mays L.) Cultivation in Ghanaian Agriculture. 2009.

Band 16 Broder Breckling / Richard Verhoeven (eds.): Large-area effects of GM-Crop Cultivation. Proceedings of the Second GMLS-Conference 2010 in Bremen. 2010.

Band 17 Broder Breckling / Richard Verhoeven (eds.): GM-Crop Cultivation – Ecological Effects on a Landscape Scale. Proceedings of the Third GMLS Conference 2012 in Bremen. 2013.

www.peterlang.de